## THE MURDERER
## AND THE GIRL

*She looked fourteen or fifteen to me. A pretty little brunet. I parked the car in an alley at the back of some shops.*

Coming off the bridge, she saw a shadow figure pacing up and down. She ignored him and tried to walk past. His arm shot out and hooked around her neck. She started to scream. She felt the blade of a screwdriver against her neck. He whispered, "Shut up or I'll *kill* you! It'll all be over in a bit." He shoved her down, letting her feel the blade while he unzipped.

*She never shouted. She never screamed. I told her she was stupid to be walking home alone at that time of night. And where the bleedin' hell were her parents? And why didn't they come to fetch her? "Let this be a lesson," I told her. "You never walk the streets at night. . . ."*

## THE MURDERER'S GUILT
## WAS WRITTEN IN HIS OWN BLOOD.
## WHO WAS IT?

THE BRAWLING EX-SOLDIER . . . one of the victims was his pretty stepdaughter

THE ORANGE-HAIRED PUNK . . . he'd been spotted near the crime scenes

THE KITCHEN PORTER . . . young girls complained of his sexual attentions

THE BAKER'S ASSISTANT . . . he had been in trouble before

THE PIPE LAYER . . . he loved the ladies and skipped the country

IF IT WERE NOT FOR ONE ASTONISHING
GENETIC BREAKTHOUGH, THIS RAPIST-
MURDERER WOULD STILL BE LOOSE TODAY. . . .

# THE
# BLOODING

# THE
# BLOODING

## Joseph Wambaugh

A PERIGORD PRESS BOOK

BANTAM BOOKS
NEW YORK · TORONTO · LONDON · SYDNEY · AUCKLAND

*This edition contains the complete text*
*of the original hardcover edition.*
NOT ONE WORD HAS BEEN OMITTED.

THE BLOODING

*A Bantam Book / published by arrangement with*
*Perigord Press / William Morrow and Company, Inc.*

PUBLISHING HISTORY

*Morrow / Perigord Press edition published February 1989*

*"Phantom Life," by Arthur Upson, reprinted from John Bartlett,* Familiar Quotations,
*11th edition, 1943. Copyright 1937.*

*Bantam / Perigord Press edition / December 1989*

ISBN 0-553-76330-X

*Published simultaneously in the United States and Canada*

Bantam Books are published by Bantam Books, a division of Random House, Inc. Its trademark, consisting of the words "Bantam Books" and the portrayal of a rooster, is Registered in U.S. Patent and Trademark Office and in other countries. Marca Registrada. Bantam Books, 1540 Broadway, New York, New York 10036.

PRINTED IN THE UNITED STATES OF AMERICA

*For my father,
with appreciation for
the genetic fingerprint*

# AUTHOR'S NOTE

This is the true story of the Narborough Murder Enquiry, the world's first murder case to be resolved by "genetic fingerprinting," a stunning scientific discovery that may well revolutionize forensic science as dramatically as fingerprinting did in the 19th century.

As always, I have re-created events only when my information comes from a reliable witness or can be independently corroborated.

# ACKNOWLEDGMENTS

Many thanks are offered to scores of people in the villages of Narborough, Littlethorpe and Enderby who offered cooperation and kindness. And much gratitude is extended to the officers of the Leicestershire Constabulary, particularly Detective Chief Superintendent David Baker.

*My days are phantom days, each one*
   *The shadow of a hope;*
*My real life never was begun*
*Nor any of my real deeds done.*
                    —ARTHUR UPSON, 1877–1908,
                         *"Phantom Life"*

# 1
# Three Villages

They say that in remote little English villages a newcomer can be accepted by the locals provided he buys property, pays his bills, and stays in continuous residence for about ninety-five years. The village of Narborough isn't remote, being only six miles southwest of the city of Leicester, and it's no longer so little—what with land developers enticing young families from urban housing estates with promises of safety and serenity—but it's a village nonetheless.

If you have an ear for pub chat, you might still hear an arcane village debate between elderly dart players as to whether or not Narborough existed during William the Conqueror's Domesday survey of 1086. A pre-Norman grave cover found in the garden of Narborough House dates from the 10th century, so it can be argued that Narborough was there, Domesday or not.

"If we ain't in the Domesday Book, we *wasn't*," is answered by "If we was in a Saxon graveyard, we *was*."

But the young villager of today cares less about Domesday and more about the fact that Narborough is woefully short on entertainment; the church/pub ratio is two churches, two pubs. Well, three churches if you want to count the Catholics who showed up about forty years ago.

Narborough has a chemist's shop for pharmaceutical needs, a bakery and confectionary, a tobacconist, a mini-market and a National Westminster Bank. Also,

there's a general store with an off-license, and there's
"R. H. Howe, High Class Family Butcher," whose sign
says: EST. 17TH CENTURY. That's just down the road from
the fruit and vegetable shop, which is next to the fish-
monger's, which is across from the Narborough post
office, whose lobby measures ten by fifteen feet.

And that's nearly all there is to the commercial
center—that and the two public houses. They complain
that the village is woefully short on pubs, but they've
got two "boozers" in neighboring Littlethorpe, one of
which serves very good Leicestershire pork pies.

The village of Littlethorpe is on the other side of the
river Soar, ten minutes down Station Road (as measured by walking time), just past the tiny Narborough
train station, erected in Victoria's day. Bordering
Narborough on the north, a brisk twenty-minute walk
up Ten Pound Lane, is the village of Enderby with a
different sin-and-salvation mix: *seven* boozers, two
churches. Enderby used to be a quarrying town and the
working-class villagers look upon Narborough as being a
bit upper crust, or "crusty." The recent growth of
Narborough, Littlethorpe and Enderby causes fear that
village identity will be lost entirely. Some think Enderby
is already more of a township than a village.

But despite local debate among members of the parish council about the alarming influx of strangers, to an
outsider the three communities still seem to be typical
English Midlands villages. There are reassuring granite
churches with mossy slate roofs, turrets and parapets
glowing rosy and amber in the setting sun. The church-
yards are cluttered with whimsical tottering headstones,
parted by irregular stone footpaths worn shiny through
the centuries. There are still enough whitewashed Tudor cottages with gnarled black beams and thatched

roofs two feet thick, carrying the trademarks of master thatchers.

There is still the comfort of strolling on village pavements, too narrow for passing baby prams, along streets barely wide enough for two cars. Most villagers trouble to keep their doors and gutters painted: green, red, black, blue. And if all the door knockers and letter boxes are no longer brass, the shiny steel substitutes bespeak a certain pride that might not be found in many of the housing estates in the city.

Perhaps a village is still a village as long as local publicans need to post notices that say: PLEASE KEEP ANIMALS OFF SEATS. Smoky village pubs tolerate lazy terriers and setters, and even pit bulls, which obviously have better press agents than do their American cousins. Pub dogs must have the highest rate of lung cancer in the animal world, but their presence remains as reassuring as empty milk bottles on the step of a whitewashed cottage.

Nobody gets specific as to how a Leicestershire country pub or village pub differs from those in the cities, but everyone agrees that they do. It's more than lower ceilings, adzed beams and ancient undulating floors. More than carpet, curtains and wallpaper, those touches that help define the uniquely British institution that acts as halfway house between taproom and home. Whatever it is, the village pubs *are* different. The Leicestershire pork pies taste better, and the Stilton cheese in a ploughman's lunch is, well, different. The village parish council provide parks and play areas, roadside seats, pavilions, nature boards and even a community center, but the village pubs, like the churches, provide a *resting* place.

The total population of all three villages is about twelve thousand souls, not counting the residents confined in Carlton Hayes Hospital. In 1938 the villagers thought it a good idea to change the name of the hospital, which was then called the Leicestershire and Rutland Lunatic Asylum. The hospital's farmland divides Narborough and Enderby and borders two footpaths: Ten Pound Lane on the east and The Black Pad on the west, names that came to evoke fear and dread.

The M1 motorway cuts across the eastern tip of Narborough, near the Narborough Bogs, and turns directly north through Enderby, next to the playing fields of Brockington School where Mill Lane joins Ten Pound Lane. It was long debated whether or not a scream of terror from Ten Pound Lane could be heard across the six lanes of that motorway.

But there's no debate that it's in a village pub that an outsider can often come closest to monitoring the local pulse. It was in the village pubs that reporters would meet to seek gossip and tittle-tattle during the time of the Narborough Murder Enquiry. And it was in a pub that a casual comment would lead toward the solution of a case destined to become a landmark in the annals of crime detection.

# 2

# Demon and Spirit

Leicestershire is one of England's smallest counties, formerly pastoral, now about half arable, well known for its cheeses and pork pies. It's had its moments of historical importance: In 1485 Richard III died here at Bosworth Field and thereby lost his crown to Henry Tudor. The county is divided east and west by the valley of the Soar, and the many hedgerows hearken to an ancient blood-sports tradition. In fact, there's still a fair amount of fox hunting in the county despite the relentless interference of animal rights groups. The devotees fear the day when one of the grannies with a sign saying TAKE THIS OLD FOX INSTEAD! gets accidentally trampled flat by horse hooves in some ditch beside a hedgerow.

Just a fifteen-minute drive northeast from Narborough on Leicester Road is the city of Leicester, county seat of Leicestershire. It's long been the industrial center of the county, the traditional industry being hosiery and footwear. Nowadays, engineering, plastics and printing industries prosper there. It's a city of nearly three hundred thousand, and, like most of Britain, has acquired a large Asian and East Indian population. The city as well as the county is policed by the Leicestershire Constabulary, a force of more than seventeen hundred officers.

The dialect of Leicester, particularly as heard around the working-class estates of the city, requires a bit of an ear.

For example, "Gizza looka 'at" translates to "Give me a look at that."

If a pub customer asks for a particular beer that's not out front he may be told: "You gorra way or else *you* gorra ge' a bokkle from ow a back." Which means "You've got to wait or else *you've* got to get a bottle from out back."

"Toym ago" means "Time to go."

"Flippin 'eck" is "flipping heck," heard almost as often as "bloody 'ell."

Verbs sometimes lose their way. "Oh, you bleeda-ah" or "Oh, you bleeder are" means "Oh, you *are* a bleeder."

As in other parts of England, short vowels are tortured so that a word like "crux" sounds like "crooks," requiring a bit of time to adjust to profanity. At first, "Oh, you bleeda-ah!" is less perplexing than "Oh, you fooka-ah!" or "Oh, you koont-ah!"

Probably the most frequently heard expression is "m'duck," with the same meaning as "ducky" or "ducks" in London. Except that in Leicestershire is sounds like "midook."

The people of Leicester have acquired an unfair reputation for being "offhand." By that, their critics mean a bit cool and abrupt—civil, but not friendly. Yet it's hard to judge people harshly when they sprinkle their speech with endearments like "m'duck."

Moreover, Leicester natives are quick to note they are a diverse lot and shouldn't be lumped in one porridge bowl. After all, the Leicester folk point out that their city produced John Merrick and singer Engelbert Humperdinck. "We had the lot," they say. "From the Elephant Man to the Wolf Man."

All in all, it's quite an agreeable city, but with a short commute to a countryside in which there are still pastoral scenes worthy of John Constable, it's no wonder that many city dwellers hope to raise their children in the villages.

Living in a Leicester working-class housing estate in the winter of 1983 was a thirteen-year-old girl about to experience first romance. She was having more trouble with adolescence than most. As an adopted child, she had difficulty with the idea of it. At a later time, she was *very* quick to point out to police officers that she'd been adopted by her parents.

"They were too strict and protective, me adopted parents" is how she put it to detectives investigating murder.

"Dad got keen on CB radio," she told them, "and I were bored most of the time. The CB gave me something to do. I got to talk to all sorts. Got chatting with this bloke."

He came to her as Spirit, his CB radio name.

Her CB handle was Green Demon, and very soon Green Demon and Spirit exchanged personal information. Spirit was fourteen, only nine months older than Green Demon, but he didn't want her to know that. A big lad, he reckoned he could fool her, and opted for more worldliness. He told her he was fifteen.

Green Demon might have settled for a radio romance, but Spirit wanted an "eyeball" and so they decided to meet at Rowley Fields School in Leicester.

Demon wasn't all that disappointed in Spirit. He wasn't dreamy but he wasn't bad—a scruffy boy, with rosy cheeks and untidy umber-brown hair and grimy fingernails. She later said he always wore T-shirts and a greasy denim jacket and jeans "covered with spots" from grease and oil stains. He was crazy about bicycles and couldn't wait to own his own motorbike. A keen mechanic, he professed to know a lot about engines.

Her menstrual periods hadn't begun yet, and she'd had no sexual experience. Nor had Spirit, but he was eager. After some rough preliminaries, a bit of petting and fondling—after he was just begging to do it—one spring night in Jubilee Park, near the Foxhunter Roundabout in Enderby, she decided to let him.

Of course a deflowering seldom makes the earth move, but Demon's was particularly unpleasant. What she couldn't forget was how he looked when he was inside her for the very first time. He just stared. He hardly uttered a sound. His brown eyes just *stared*.

It was unsettling for the girl, and if that wasn't bad enough, when it was over they sat and chatted about records, and motorbikes, and *anything* other than the Great Moment they'd just experienced.

She later said: "It were not mentioned! Just as though it hadn't happened!"

Spirit's appetite had been tapped by Green Demon. There was sex whenever he could get to her, but during it he never took off an article of his own clothing, and she was permitted nudity only from the waist down. She started her periods later that year, and the encounters continued into the autumn when she began to worry about pregnancy. They never used contraceptives and Demon had some thoughts about what her parents might do if she came up pregnant and she not fourteen until December.

She started resisting Spirit's intimacies, but he'd get furious when she did. He'd swear and call her names. Spirit was growing bigger and stronger and he'd grab her shoulders and shake her. He even slapped her across the face and they had sex while she wept, humiliated.

When his parents weren't home she went to his house for bedroom sex. Whenever she refused him, he punched her in the stomach and forced her to the floor with his hand on her throat. Even when he got his way, he called her a "slag."

Another time in his bedroom he punched her and the force bashed her head against the wall. She lay dazed and he put something foul-smelling under her nose to revive her. His younger brother, often at home when they were in the bedroom, came running.

"Ye don't know your own strength!" the younger brother cried.

Still, she returned to him. Demon was a *very* lonely girl.

Finally, first love or not, Demon began to understand that Spirit's sexual demands could get a bit dangerous, especially when problems occurred. Spirit, it seems, was suffering from "brewer's droop," as he called it. She laughed at first, but he didn't think premature ejaculation was funny. When it happened he'd rage and bang the wall with his fist, and all but weep. She'd slip away and leave him alone until the fury passed.

"Nobody likes me," Spirit would cry out. "Nobody! Especially you slags!"

Perhaps to counter the brewer's droop, Spirit started experimenting. He wanted her to "do the sixty-nine," but she thought it was immoral.

"Okay, then, let me bum you," he suggested one day, but Demon refused.

"But he just *insisted* on sex in me back passage," she later explained, "and finally I gave in."

The back passage sex, with her on all fours, began in the autumn of 1983 at the start of school term. There were some very cold days, especially in November, but according to his later testimony, the first time they

tried it was outdoors on a railway bank. Inclement
weather couldn't dampen *this* Spirit.

· She later emphasized that anal sex became as frequent as vaginal sex, but even that wasn't enough. He
liked to bite. Aggressively. She didn't fancy all those
love bites on the neck and shoulders and on the inside
of her thigh; it was almost as bad as being backpassaged.
Finally, Spirit's sex play got *too* rough.

"Once when we was in his bedroom he began mess-
ing about with belts," she said. "He tied up me hands
and pulled off me knickers and did it to me."

That did it for her. Demon met another boy and the
contrast opened her eyes. She rang up and said it was
over and she never wanted to see him again. He called
her a slag.

As the girl matured, and was forced to remember her
sexual experience with Spirit, she recalled that it had
always been the same as the very first time. Afterward,
he'd play records, or talk about motorcycles, or cars, or
make other small talk.

"We'd simply carry on as if nothing ever happened.
Nothing at all."

It was the strangest thing about him, she decided. It
was even more weird and disturbing than being bound
helpless while he *stared*.

Spirit later told his own version of the brewer's
droop humiliation that had plagued those first encounters.

"She were laughing at me when it happened," he
later confided. "They *always* laugh at me!"

"Who?" he was asked by his interrogator.

"Them!" he answered. "Them. I call them slags,
dogs, whores, bitches. *All* of them."

# 3
# The Black Pad

One of the city dwellers who thought village life would be healthier for her two daughters was Kathleen Mann. Kath was Leicester born and raised, and when her five-year marriage ended in 1970 she brought her tots to her mother's house in the city. But Kath quickly learned that two mothers make the job twice as hard. After enough disputes over child raising, she arranged with her brother to take his subsidized flat in the village of Narborough when he moved out.

To Kath, Narborough was all that an English village should be. You could go for lovely strolls down Church Lane, past cottages with bottle-glass leaded windows, past ancient doorways framed in a time when robust country squires seldom topped five feet three inches in height. It was fun to watch tall young villagers passing in and out of cottage doors, in a semi-genuflection.

Nearly everyone had a garden. There were smells of wood smoke and carnations, and climbing roses on trellises. You might spot a treat almost anywhere, such as Victorian birdhouses with individual rooms and perches—solidly built and painted a hundred times over the years—nearly as eternal as the oaks in which they rested. And just beyond the winding village streets, sheep and cows grazed in summer pastures under oatmeal clouds.

"A typical English village," Kath called it.

She was head of her own household and her children

were village children, out of the city, out of harm's way. But village life was not all teatime and violets, not by a long shot. Her subsidized home was actually just a cold-water flat with an outside toilet, and besides, there was a void. Kath Mann was a woman alone with two daughters for nine long years.

Then she met Edward Eastwood in a singles club at the Braunstone Hotel in Leicester. He was nothing whatever like her first husband, and nothing like herself. Kath was a short, buxom brunet, serious and shy. Eddie was a strapping, fair-haired talker with big expressive hands. His hair had a kind of curl, and with his horn-rimmed glasses, he might (*if* you fell in love with him) seem a roughhewn version of a Michael Caine Cockney. And Eddie Eastwood from Yorkshire was as glib as any East Londoner born within earshot of the Bow bells.

He was the kind of bloke with a hundred tales, all of them colorful, replete with hyperbole. An ex-soldier, he regaled his listeners with stories of barroom brawls, and even claimed to have been shot by an Arab terrorist. No one knew what to believe when it came to Eddie Eastwood—who actually had changed his surname in a court of law—which, they said, probably made Eddie Eastwood the leading fan of steely-eyed Clint in all the British Isles.

Eddie seemed a hard man, but it was mostly bluff. He'd lived his early years in the Braunstone Estates, called "Dodge City" or "The Badlands" by local police. They said he had had some rough chums in the old days, but he was an outgoing friendly sort, and passed many an hour in the pubs playing darts and drinking bitter. Eddie Eastwood was easy to like, and had a reputation as "a good pub mate."

In July, 1980, when Susan Mann was fourteen and

Lynda was twelve, Eddie Eastwood moved in. He married Kath in December, and took his new family to a semi-detached home near Forest Road, by the psychiatric hospital. By The Black Pad footpath.

Things went very well for the Eastwoods. Eddie earned a fair wage and they lived in a street called The Coppice. They had a small greenhouse, and Eddie built a huge aviary from packing cases. He got so interested in raising and caring for budgerigars that he built yet another aviary, and ended up with seventy-two of the parakeets, along with dogs, cats and guinea pigs for the kids.

Eddie was working ninety hours a week at Spray-Rite Ltd., "paying double whack" to clear up old bankruptcy debts, yet in his spare time he won trophies, both with his budgies and his darts. The following year, when Eddie was thirty-nine and Kath was thirty-three, they had a baby girl and named her Rebecca. Those years were the best they would ever have together.

Of Kath's older children, Susan was the shy one. "She was a home girl," her mother said. "Much like me, I think."

Susan had dark-blond hair and eyes as quiet as a spaniel's. She liked to stay at home and play with the animals and birds. She wasn't as attractive or as bright as her younger sister, and seemed aware of it.

Lynda Mann had little trouble with adolescence. An adventurous girl, she liked *everything* about growing up: music, hairstyles, makeup, clothes. She got her share of A's and her headmaster at Lutterworth School was pleased with her. When she turned fifteen Lynda talked about being a multilinguist, and practiced her French, German and Italian. She wanted to try Chinese, announcing that she would one day travel the entire world. Her mother didn't doubt for a minute that

Lynda would do whatever she set out to do. If money for clothes was a bit short, never mind, Lynda would babysit and earn money and make her *own* dresses.

There were a few boys, one in particular, but Lynda was fancy free in 1983, a fifth-former at Lutterworth. She was particularly upbeat that term, growing more fetching every day. Lynda's hair and eyes were very dark but her skin was fair. People referred to the fifteen-year-old as "happy-go-lucky" or "bubbly" or, at the very least, "enthusiastic."

The weather in November turned bitter and bleak. Monday, the 21st, was predicted to be very cold. When Lynda Mann dressed for school that day she wore tights, form-fitting blue denim jeans with zips at the ankles, a pullover, white socks and black tennis shoes. Before setting foot out the door, she snuggled inside her new donkey jacket with the stand-up collar, and, for good measure, put a warm woolen scarf in her pocket. She walked to Desford Road and traveled to Lutterworth by bus.

Lynda came home that afternoon on the school bus, but had no time for "swotting," cramming for exams. She had to babysit at 5:00 P.M. for a neighbor near the Copt Oak housing estate. She babysat until 6:20 P.M.

Supper for Lynda meant plenty of salad cream, which she dumped on practically anything. She and Eddie had a quick meal together, then she changed into a mauve sweatshirt and was off again at 6:45 for another babysitting appointment at the home of Mrs. Walker, a nearby neighbor, who was waiting outside her house when Lynda arrived.

"Sorry, dear," Mrs. Walker told her, "I'm on sick leave from work and won't be needing you today."

Lynda was disappointed, but shrugged, smiled, and

said, "Well then, I'll just go home, or maybe to Enderby to see a friend. Bye!"

"Bye, dear," Mrs. Walker said.

It was 6:55 P.M. By the time Lynda got home a full moon had risen. A blanket of frost had settled on the ground in the Eastwoods' garden. Lynda told Kath she wouldn't be earning any money that night, then said she'd be going to visit her best friend, Karen Blackwell. Lynda had £1.50 she'd saved from babysitting to pay toward the donkey jacket, which she'd ordered from a shop-at-home catalogue that Mrs. Blackwell had shown her.

"Are you coming straight home?" her mother asked.

"I'll probably stay at Karen's for a while, then I might just stop to see Caroline," Lynda said. "Don't worry, I'll be home by ten."

"Independent," her mother always said, when discussing Lynda. "The child is *so* independent."

She was the sort of girl who didn't want much parenting. Lynda seemed to know exactly where she was going in life and performed so well in so many ways it was hard to bridle this middle child. If she wanted to dye her brown hair a darker shade, well, what could you say? It was better than the henna red still showing from her previous experiment.

There was an occasional nagging worry for her mother. Lynda had had a steady boyfriend during the prior year and a couple of casual ones, and being she was so young it caused Kath a bit of concern. And Lynda had met another boy at the Lutterworth School disco, a boy Eddie called a half-caste. Yet as far as Kath knew, her daughter did not smoke or drink, and Kath believed her daughter to be a virgin. Eddie often said that Lynda was nobody's fool, and so a mother needn't worry too much.

Shortly after 7:00 Lynda walked down Redhill Avenue, not the most direct route to Karen Blackwell's. She was seen by a friend named Margaret, who asked where she was going.

"Down a friend's," Lynda replied.

Margaret later said that Lynda "was her normal cheerful self."

A few minutes later Lynda arrived at the Blackwells'. Karen had been Lynda's best friend for seven months, and they'd known each other since primary school and at Lutterworth. They were in different classes with different teachers, but were the same age and shared adolescent confidences.

Lynda gave the £1.50 to Mrs. Blackwell who in turn would give it to an agent for Kay's Catalogue Club. Mrs. Blackwell signed the club card, reporting the payment toward the donkey jacket.

The Blackwells liked and approved of Lynda Mann. "A quietly spoken, well-mannered young lady," Mrs. Blackwell said of her daughter's best friend.

Then Lynda said, "Well, I'm off to Caroline's to collect a record I've loaned her."

Caroline lived in Enderby, about a fifteen-minute walk from the Blackwells', up Forest Road, near The Black Pad.

"I knew it was about half seven," Caroline later said, "because Lynda was in and out the door before the music for *Coronation Street* came on the telly."

Lynda walked up Forest Road, toward the streetlight where a footpath leads off toward the pastureland belonging to the psychiatric hospital and joins The Black Pad, the lonely path that angles down toward the cemetery behind Narborough church.

Lynda saw a figure standing by the lamppost. He had placed himself in the light like an actor on his mark. He was not far from the gate of the Carlton Hayes psychiatric hospital. On that gate was a sign warning motorists who might enter through the gateway. The sign said: DEAD SLOW!

———

Kath and Eddie Eastwood had themselves a pleasant evening. First they attended a ladies' dart tournament at the Carlton Hayes Social Club. Then they were off to The Dog and Gun, a favorite pub of Eddie's where he managed to win a few pints of bitter playing darts until 12:10 A.M. One of his victims was a local policeman, which evoked the expected jokes about getting back some of the taxpayers' money from the coppers.

The Eastwoods arrived home about 12:30 A.M. and found Susan waiting up.

"Lynda's not home!" Susan said.

Eddie Eastwood drove around village streets and checked teenage gathering spots. One of the places Eddie searched on foot was The Black Pad, near the Eastwood home. They were building a new housing estate of upmarket homes on one side of the footpath, opposite the psychiatric hospital's pasture. The workers already had the foundation poured and lumber stacked, but hadn't done much framing.

Eddie walked the length of the unlit Black Pad, alongside the housing development. It was then that he noticed how really bright and clear it was. Walking The Black Pad at night was usually a bit unnerving, and the moonlight helped.

Eddie called the Braunstone Police Station at 1:30 A.M. to report Lynda missing. A policeman took down the information, but policemen the world over don't get very worked up about fifteen-year-olds a few hours late.

"But she's always home by ha' past nine," Eddie told
the officer. "Unless we know, she's *always* here!"

When Eddie had searched The Black Pad, it seemed
logical to him to look toward the side where the new
construction was under way. If there were any teenagers up to mischief, or, God forbid, if anything bad had
happened in that dark lane, he'd find evidence there by
some lumber pile, he thought. The other side of the
footpath was protected by a wrought-iron fence more
than five feet tall, a permanent barrier separating The
Black Pad from the grounds of the psychiatric hospital.
Near the top of the stanchions the black iron bars
curved toward the footpath like a row of iron claws,
menacing those who walked The Black Pad.

He had seen nothing move, and heard nothing except the tree limbs, bare of foliage. They groaned in the
wind under a blue-black sky, a glittering moon, a few
shredded clouds. Edward Eastwood had never thought
to look toward the hospital side as he picked his way
through the darkness down the black tarmac footpath.
He had passed within a few yards of his stepdaughter,
Lynda Mann.

# 4

## Mannequin

A hospital porter who often used The Black Pad as a shortcut between Narborough church and Carlton Hayes Hospital was on his way to work at 7:20 on Tuesday morning, November 22nd, when he glanced through the wrought-iron fence, toward the wooded copse and grassy fields of the hospital grounds, white with frost on that cold morning. He saw what looked like a partly clothed mannequin lying in the grass by a clump of trees. He stopped and gaped. She was naked from the waist down. There was a smear of red about her nose. He was not sure if she was real.

The porter ran out of The Black Pad onto the road and flagged down a car driven by a colleague, an ambulance driver from the hospital. The ambulance driver and the porter jogged back to The Black Pad and looked through the fence.

"Is it a dummy?" the porter asked.

The ambulance driver ran to the head of the path and found the iron gate wide open. He entered the grassy field and approached. Lynda Mann's jeans, tights, underpants and shoes were in a rolled-up heap about ten or fifteen feet away. Her legs were extended straight out, her head turned to the right. She was supine with the upper part of the donkey jacket hiked under her head, the sleeves partly pulled up her arms. Her chin was bruised and there was bright coagulated blood from her nose. Her scarf was wrapped around her neck and

27

crossed at the back, and a piece of wood about three feet long lay under her right leg.

Perhaps the ambulance driver was familiar only with victims very much alive and breathing, including those who screamed and thrashed inside straitjackets. Maybe he felt the need to display medical training in the presence of the porter. For whatever reason, he reached down and felt the throat for a pulse, even though rigor was present throughout.

Lynda Mann was white as china. As rigid and cool as a shop mannequin.

---

It had been an unforgettable year for the Leicestershire Constabulary. The county police agency averaged about one homicide a year and usually that was a domestic killing. But that year had seen four murder inquiries, two of them major, culminating in the tragic discovery in July of the body of five-year-old Caroline Hogg, who'd disappeared from a fun fair near her home in Edinburgh.

The Leicestershire police always believed that the child's killer had arbitrarily dropped her body by the A444 road while passing through from Scotland to some southerly destination, but because they'd found the body, they had to launch an inquiry from their end.

Detective Superintendent Ian Coutts, born and reared near Glasgow, went up to Scotland for assistance with the Hogg case, and to gain access to the Edinburgh computer. The fifty-year-old Coutts was a "typical Glaswegian": gregarious, outgoing, tough, solid and compactly built. It wasn't hard to imagine broad foreheads like his greeting adversaries with a "Glasgow kiss," the kind that leaves many a bloody nose in northern pub brawls.

It took an enormous amount of work to back-record and convert material that had to be manually accessed with the Leicestershire card index system.

Then there had been the Osborne murder, the case of a pet groomer brutally stabbed to death and left on Ayelstone Meadows. That one had required a scene-of-crime fingertip search for evidence in ferocious driving rain. They'd remember *that* one. On the Osborne inquiry they'd had to access a West Yorkshire computer and put their material into it. Until that terrible year they'd always had sufficient data-processing capability in their own computer terminals.

There was a joke making the rounds of the Leicestershire Constabulary that year: "Did you hear the good news? Yuri Andropov died. The bad news is they dropped his body in Leicestershire."

But until November of 1983 there had never even *been* a murder inquiry in the villages of Narborough, Enderby and Littlethorpe.

The detective chief superintendent in charge of Leicestershire Criminal Investigation Department was forty-seven-year-old David Baker, a twenty-seven-year police veteran. Baker was a family man with an accommodating style. He looked more like an avuncular shopkeeper than a policeman, but he was, in the words of close associates, "one hundred percent copper." He had five kids, and managed a squash game at least once a week in a losing battle with middle-age spread.

At 8:30 A.M. Chief Supt. Baker arrived in Narborough, logging his location as "a wooded copse running alongside a footpath known as The Black Pad." There were many police officers already at the scene, and Baker called at once for a Home Office pathologist. The Lynda Mann murder inquiry had officially begun.

Several detectives, and thirty uniformed officers along

with tracking dogs, began searching the copse, the fields, the building site by the footpath, and The Black Pad itself. When the pathologist arrived he made notes: that rigor was present, that there was blood showing at the nostrils, that there were scratch marks on the upper right cheek and below the right orbit, that the tip of the tongue was protruding through the clenched teeth of the strangled girl. The police had thought that her legs were painted with some sort of brick-colored leg makeup, but learned from the pathologist that extreme cold had produced the effect.

The pathologist noted that there appeared to be "matted seminal stains on the vulval hairs."

———————

For the Eastwoods the memory of the day would forever be hazy. Eddie went to his job and told his workmates that his daughter was missing. When he was informed about a body having been discovered alongside the footpath he left work and drove straight to The Black Pad, finding the area cordoned off by several bobbies and detectives already trying to organize a house-to-house inquiry.

Eddie tried to push through a barricade, but was stopped by one of the policemen.

"That's my daughter!" he told them. "I *think* that's my daughter, Lynda!"

The policeman began talking to him and making notes while Eddie Eastwood had the thought all frightened survivors have at such a moment: Well, of course! It's all a mistake!

He later remembered the policeman saying, "Go home. We'll call on you."

*    *    *

The rank of inspector in the British police service is the equivalent of lieutenant in most American police forces. One of the CID inspectors, Derek Pearce, had just come off the aborted Caroline Hogg inquiry. And Pearce absolutely hated leaving an inquiry "undetected."

Derek Pearce was the kind of whom they say, "You either like him or you don't." They also say that Pearce had the ability to rise to the top of the police totem if only he weren't constantly being dragged back down *by* Derek Pearce.

Members of the Lynda Mann murder squad asked to name the brightest detective among them responded:

"Derek Pearce."

"Oh, Derek Pearce, of course."

"Pearce, no doubt."

"Derek Pearce, *but* . . ."

There would always be a "but" with a man like Derek Pearce. Some of the adjectives preceding his name were: immature, talented, abrasive, ruthless, charming, insensitive, generous. But *everyone* called him complex. A driven perfectionist, he expected everyone to do the job as well as he would.

To get an idea of his energy you'd only have to watch him for an hour. If he was on his feet talking to someone he'd rise on his toes, or rock back on his heels, or slide, or bounce, or sway. If he pursued his listener through a doorway he'd stop, grab each side of the jamb, and do what looked like calisthenics or yoga: pushing, pulling, rising, settling. They said if you could harness that energy you could power British Rail.

He did everything at his own pace, from driving a car (daringly) to mixing beer, Scotch and vodka (daringly). And Derek Pearce fed on stress. If the job didn't supply enough anxiety, he'd find some. John McEnroe would understand.

Pearce was thirty-three years old, just under six feet tall, and slim. But he seemed *very* slender. Anyone who survives such an energy overload *seems* very slender. In a pin-striped suit he could've been a young barrister at Crown Court. His thatch of darkest-brown hair and Royal Shakespeare Company beard were closely cropped. He could've played Petruchio in *Kiss Me Kate*.

Pearce's glasses made his brown eyes dilate when he was flying into someone's face. Occasionally he was imprudent enough to take intensity-fueled flights at superior officers.

His four-year, childless marriage to a policewoman had just ended, and Pearce lived alone with a very large English sheepdog named Ollie. Police work was his life.

Pearce had been off the Hogg inquiry only one day when he received the word from a fellow detective inspector: "We've got another murder. This one's in Narborough."

"You're winding me up!" he said. "Tell me it's a joke!"

"It was a quirk of fate that I got on the case at all," he later recalled. "The other DI had to go to the Crown Court, and it isn't easy to get out of it when you're summoned to the Crown Court."

"It's all yours," he was told by the court-bound DI. "Cheers, mate."

Pearce immediately organized the call-out of a squad that would grow to 150 men and women.

When Derek Pearce got to the crime scene that day, he was told by Chief Supt. Baker to drive to the home of Edward Eastwood and bring him to identify the body which they were reasonably certain was that of his stepdaughter, reported missing the night before.

At 11:15 A.M. Eddie Eastwood was back at The Black Pad where he met Baker and a scene-of-crime officer who was trying to gather exhibits.

When a detective lifted the shiny black sheet, Eddie Eastwood looked and said, "That's Lynda."

"Are you sure?" the detective asked.

"Put it down!" Eddie said. "That's her."

Death had sculpted and painted and done its masquerade. But Eddie knew that the pale contorted face with the half-lidded wistful gaze and clenched tongue belonged to his pretty stepdaughter, Lynda Mann. He knew because he recognized the donkey jacket that she'd been paying for a bit at a time.

---

Pearce was allocated to deal with the family. He took with him Detective Sergeant Mick Mason, a perfect choice. The family immediately took to Mason.

"Mick's that kind of bloke," Pearce said. "One hundred percent solid."

Mick Mason promised the Eastwoods he'd keep them informed of the progress of the inquiry until it was over. Like the others, he thought it might last a week at most.

Kath Eastwood was, as she later put it, "in a daze" that day. She sat at home and offered tea to the various detectives who came.

"I had ever so much trouble," she later remembered, "just trying to understand anything said to me."

Did Lynda have a boyfriend? How about a former boyfriend? Lynda's ears were pierced but she wasn't wearing earrings. Was she wearing earrings the last time you saw her? Questions like that.

That day, Kathleen Eastwood sat and tried to recall whether or not Lynda had been wearing her ear studs.

*The last time you saw her.* It took a few minutes even to begin to conceptualize what that must mean.

"She might've been wearing ear studs" was all Kath could offer. "And gloves. Perhaps she had her gloves, ones without fingertips."

But she couldn't be sure. Everything was mixed up. This couldn't be happening because Lynda had planned her future so completely! It was all Kath could think about. Lynda's future was assured.

*The last time I saw her.*

# 5
# Victims

A postmortem was conducted at 2:00 P.M. on the 23rd of November at the Leicester Royal Infirmary with Chief Supt. Baker and Supt. Coutts in attendance.

The body of Lynda Mann measured five feet two inches. It weighed 112 pounds. Facial abrasions were noted on the right upper cheek and orbit, as well as on the chin and the front of her neck. Detectives speculated that she may have been punched on the chin, perhaps knocked unconscious by her assailant. There was bruising below the middle of the left and right clavicles.

As to those chest bruises the pathologist wrote: "Conditions would suggest that this girl was struck a heavy blow to the upper chest." Detectives, however, speculated that the bruising might have been caused by an assailant kneeling on the girl to provide leverage as he tightened her own scarf around her neck as a ligature. By the left knee if he was right-handed.

The fingernails were long, partly painted and unbroken, indicating that Lynda Mann probably had not put up a terrific fight. There was no damage to the anus, and the vagina showed no tears or bruises. The tongue had been gouged by her own teeth as she died strangling. She was probably not unconscious when the final force was applied.

The principal scientific officer from the Forensic Science Laboratory at Huntingdon doubted there had been a protracted struggle. He wrote that the arrangement of the tennis shoes "indicated voluntary removal." He found the pubic hair around the vagina to be matted with dried fluid, probably semen. Soil marks on the back of her bare heels suggested that her body had been dragged by the upper half, perhaps by the donkey jacket.

It was discovered that the zip tab on her jeans was jammed, consistent with the belief that they had been stripped from her body by the assailant who had left them rolled up, inside out. Officially, the cause of death was from asphyxia due to strangulation.

At the conclusion of the postmortem report the pathologist wrote: "Sexual intercourse was attempted and premature ejaculation occurred."

A worried spokesman for Carlton Hayes Hospital gave a statement to inquiring reporters: "Although this tragedy is right on our doorstep, there is no reason to suspect there is any connection with the hospital."

Which, when reported to Derek Pearce, caused him to look at the looming psychiatric hospital and say, "Oh yeah. Loonies and maniacs hanging out every flippin window and there's no reason to suspect any connection!"

Eddie Eastwood gave the interviews for the family when reporters from all the media arrived at the little semi-detached council house in Narborough.

Eddie sat, head in hands, and said, "I feel like I've been hit over the skull with a brick. It just will *not* sink in. Lynda was such a happy, polite girl. A little old fashioned but very popular at school and conscientious about her schoolwork. It's unbelievable that anyone could do this to her. My wife and I are devastated."

Friends of Kath Eastwood never doubted that. Kath just sat numbed, responding when she must. She could not even weep.

* * *

The laboratory analysis offered the murder squad its first positive clue. The report showed that the killer had, after ejaculating prematurely, managed to penetrate the victim prior to death. Semen was taken from an internal labial swab and on a deep vaginal swab.

Given a phosphoglucomutase (PGM) grouping test, the semen showed strong PGM 1 + enzyme reaction. It was antigen-tested and found to contain strong amounts of Group A secretor substance.

The officers on the murder inquiry were told that only one out of ten male adults in England was in this particular blood group. The scientific label would remain the only clue they possessed. The killer was a Group A secretor, PGM 1 +. Without understanding exactly what it meant, hundreds of police officers would repeat it for nearly four years: "We're looking for a PGM one-plus, 'A' secretor." Though a blood test could not positively identify a killer, it could be used as a tool, and it was so used on Edward Eastwood.

Eddie always believed he'd been fingered by one of Lynda's former boyfriends who had a grudge against him. Actually, Derek Pearce was following the time-worn police procedure of looking from the inside out. Eddie Eastwood was only a stepfather and had belonged to Lynda Mann's family for a relatively short time. Pearce supposed Eddie could have killed her at home and dumped her there by The Black Pad.

"I've told Mister Coutts I want to nick Eddie," Derek Pearce explained matter-of-factly to his subordinates. "And I'm going to do it."

Pearce got Eddie out of bed at nine o'clock at night. Eddie complained that he was ill, but Pearce wasn't using his bedside manner. "He's the kind that gets sick at the drop of a hat," Pearce said of Eddie Eastwood. "He's always on sick leave from work."

"I never liked him from the first," Eddie later said of Pearce. "And me stepdaughter Susan, she came to *despise* him. I heard the other CID man saying to Pearce, 'Eddie Eastwood never murdered nobody!' but Pearce said, 'I don't care, we're going to blood-test him anyway.' A *bully* copper, that's Pearce."

"Everyone else that's closely associated with Lynda is having to give blood," Pearce said to Eddie that night. "And one or two people think *you* might be involved."

"You saying I killed me own daughter?" Eddie challenged.

"*Step*daughter. No, I personally don't think so, but that's all right, I'll take you in the car and bring you back home. This'll eliminate you completely."

"I'm a sick man!" Eddie told him. "I got arthritis in the lower back!"

"Yeah, well, I'm not all that fit, m'self," Pearce said. "We'll *both* feel better when this is over, won't we, mate?"

While Eddie was shuffling past a queue of suspects in the police station that night, it occurred to him that he had actually been in the company of a policeman on the night of the murder.

"I just remembered!" he said. "I was playing darts in a pub with a copper! You can check."

But Pearce didn't seem to care if Eddie was a darts mate of the Lord Chief Justice. He took him to a doctor who drew blood from his arm. The doctor also took hair samples from his head, underarms and groin before Eddie got to return to his bed.

But Eddie's alibi checked out, and, finally, it turned out that Eddie was not in the same 10 percent blood group as the slayer of Lynda Mann.

Derek Pearce, who wore eyeglasses in his everyday life, learned something of significance to him: Lynda Mann had had poor eyesight. She couldn't recognize a friend across the street, but adolescent vanity prevented her from wearing glasses, and contact lenses were too expensive for her family.

Supt. Coutts remained convinced that Lynda Mann *must* have known her attacker, but then Coutts didn't have to wear glasses to see across the street.

Pearce wondered if Lynda could have *thought* she knew the man, perhaps walked right up to him before realizing he was a stranger. When it was too late.

---

For a man like Eddie Eastwood, finding himself on the news at ten, given the attention of the national media, it was understandable if he indulged himself a bit. Eddie later said, "The humiliation of the blood test affected me speech. I couldn't give a television interview for three weeks."

It might be impossible for a cynical policeman to believe that Edward Eastwood could be experiencing anything like the overwhelming grief that Kath suffered, but on the other hand, Eddie *had* gone looking for his stepdaughter on that bitter night. Eddie had searched the streets and walked The Black Pad under a bright and brittle moon, and it was he who had to view the blood-blackened, ruined body of Lynda Mann.

They scoffed at the many interviews in which he, as the family spokesman, uttered a litany of personal heartbreak while Kath remained silent, stoic, shattered. But perhaps after the permanence of death was absorbed— after the media attention waned, after he had been taken from his bed and made to prove that he hadn't raped and murdered Kath's middle child—perhaps af-

ter all that, even a policeman could believe Eddie
Eastwood when he said, "I went to a pub in Enderby
one day. I went into the back room and just let go. I
realized how much we were all victims of the one that
done it. I cried like a child, I did."

# 6
# Village of Fear

As darkness fell on country lanes and village footpaths, women and girls rushed to their homes. Many parents insisted on walking or driving their children to school, and some threatened to keep them home until the "fiend" was caught. There were parents waiting at bus stops for weeks after the murder. Villagers in shops and pubs spoke in whispers and eyed one another strangely.

Rumors spread about a flasher who had exposed himself to another girl on The Black Pad. And a woman claimed to have once been assaulted on a different village footpath. A third told of having been "mugged," of having her purse stolen weeks earlier on The Black Pad. None of these crimes had previously been reported.

There were calls to the Narborough Parish Council from terrified parents who wanted to light The Black Pad or close it down completely. And because a length of wood was found under the leg of the murdered girl, gossip had it that she'd been beaten half to death with a club before being raped.

The newspaper headlines referred to Narborough as VILLAGE OF FEAR. So it wasn't just foul weather that left the village lanes so bereft of nighttime foot traffic. Not just winter mist and creeping wisps of fog that made women quicken their steps, under an oppressive stalking sky.

---

Chief Supt. Baker decided that it would be most convenient if the incident room could be set up in the village. He requested assistance from Carlton Hayes Hospital and was offered a doctors' residence that the hospital hadn't been using for everyday business. The building was called The Rosings, and the commemorative stone over the door said: A.D. 1906. Most of the brick buildings in the hospital complex had been built at that time, and The Rosings hadn't been remodeled since.

The murder squad put its computers on the ground floor, and Supt. Coutts worked upstairs, along with several teams, the policewomen, and the card-index civilian workers. Other than that, they had one small room in which to relax and have a sandwich.

Computer retrieval was an art, the murder squad soon discovered. Two sergeants extracted all local handwritten records on everyone with indecency offenses, deleting those of men who were deceased or serving time in prison. They listed indecency offenders by putting rapists at the top, followed by those who'd committed rape in the general area of the Midlands, followed by rape in Leicestershire. The rape category was followed by less serious indecent assaults on females, all the way down to flashing offenses. They also computerized assaults on males, including boys, and broke these down to the flashing of males by males. Hundreds of man-hours were needed for this manual extraction before the information could be put in, and retrieved from, their new computer system. The overall plan was to link all information with names that might appear on the house-to-house inquiry.

It wasn't nearly as bad in The Rosings as it was for the hospital squad and house-to-house teams who occupied the cricket pavilion. That pavilion, across from the hospital's cricket pitch, was like a drafty shoe box built off the ground on an exposed foundation and "warmed" by an unvented gas heater. They held briefings in the pavilion, stopped there for tea and biscuits, and despite the thirty bodies working cheek to jowl it was said the pavilion stayed as cold as a lawyer's heart.

But no matter how cold and cramped it was in the pavilion or in The Rosings, they persevered, in the belief that it couldn't last long. Any one of them would've been shocked to think they'd be there at Christmas. They couldn't foresee that they'd be jogging around that cricket pitch for exercise on warm *spring* days, or that they'd be there long enough to watch the birth of daffodils in the hospital gardens, and stay to see them die.

The house-to-house teams did what the name implies— they went door to door, to *every* residence in the three villages, filling out a *pro forma* document on each male resident between the ages of thirteen and thirty-four. That age had been arbitrarily selected when it was learned from lab technicians that the sperm count in the semen sample was high. Which prompted ad-libs from the over-forty cops, such as: "Well, if I'm out of the age group, how is it I inflate the old woman every time I roll over in bed?"

Protests from middle-aged detectives notwithstanding, they investigated only younger men, and because the house-to-house teams went back five years, so did the hospital teams. It was a massive task to dig into hospital records and try to trace likely outpatients and resident patients who'd passed through Carlton Hayes over that period of time.

They had formed a hospital squad because of the number of sexual offenders, drug abusers and alcoholics treated at Carlton Hayes, not to mention the ordinary

psychotics capable of rape and murder. Hospital spokesmen were cooperative after the police offered reassur-
ances, but were understandably cautious about opening
up confidential psychiatric files. The hospital would
not give background information on patients but repeatedly assured police that the killer of Lynda Mann
could not possibly have been one of the resident
patients.

"Our wards are secure," the murder squad was told.

After which, Derek Pearce told his men, "About as
secure as Woolworth's on Saturday afternoon. You'll
just have to be resourceful and sort out as many loonies
as you can."

When he was able to assemble a true picture of the
monumental job they faced, Pearce said, "Bloody hell!
There's more people comes through this place every
day than in Euston Station at rush hour, and that doesn't
even include the day center!"

There was a little brick outbuilding across the road
from The Black Pad, on the grounds of The Woodlands,
a large hospital residence made into a day center for
people with mental problems not severe enough to be
treated in the hospital. The inquiry teams discovered
that teenagers would hang around the little brick building drinking beers or soda pop, eating sweets. They
were able to prove that Lynda had been there once or
twice with other teenagers. It was quite close to her
house, closer yet to The Black Pad. They worked the
lead exhaustively, but The Woodlands didn't seem to
figure in the murder.

Many of the day center patients had no community
ties and no family ties. They could be in Carlton Hayes
for treatment one day and arrested the next day in
Wales or Scotland. The inquiry teams were looking at

*ten thousand* hospital patients, and many of them, according to the beleaguered detectives, were potential suspects.

Almost immediately phone calls began flooding the incident room, the most promising being about a "spiky-haired youth." The person on the phone claimed to have seen him at 8:00 P.M. at the junction of Forest Road and King Edward Avenue, just a two-minute walk from the wooded copse where they found the body of Lynda Mann. The witness had been driving down the dual carriageway when the spiky-haired youth and a female companion stepped onto the road, forcing him to slam on the brakes. "The girl was wearing jeans and a donkey jacket," he said. "The young man had a dyed punk hairdo. Amazing hair. Like a pot of geraniums cropped off flat."

Within a few days after the description was reported in the *Leicester Mercury,* the police received a tip on another important suspect seen running on Kipling Drive in Enderby on the night of the murder. He couldn't have been a jogger, the caller told them. He was wearing ordinary street clothes.

Along with the reports on the spiky-haired youth and the running youth was another message given priority by Supt. Coutts. Three witnesses reported seeing a young couple in the bus shelter on Forest Road sometime after 8:00 P.M. on November 21st. The description of the girl closely matched that of Lynda Mann. It was a lead that Coutts believed corroborated the message about the spiky-haired youth in the street.

Locals told police there were *no* spiky-haired punkers in the villages, at least not one whose head resembled a flower pot full of cropped geraniums. Coutts said there was, and that they'd find him.

*   *   *

Nurses at Carlton Hayes Hospital reported being too terrified to walk from the hospital to their quarters at Sylvia Reid House, just steps from The Black Pad.

"I knew something like this would happen!" a nurse told police. "We're scared to death!"

She wanted the car park in front of their building lit, and demanded that police arrest the prowlers and vandals who came by and tossed stones at their windows.

On the eighth day they got a call from a nurse who claimed to have heard a frightened scream on the night of the murder. "A female shouted, 'No, no, no!'" she told detectives.

"There's a strong possibility that this was Lynda," Supt. Baker said to reporters. He called the lead "promising."

But Derek Pearce didn't get excited about the scream heard by the hospital nurse. The scream was timed at 8:40 P.M. and he knew Lynda Mann had left her friend at 7:26. Despite theories about the girl at the bus stop he believed that nearsighted Lynda Mann had walked to her terrible fate immediately after leaving her friend's house. Straight into an ambush.

"And besides," Pearce confidentially told his men when no bosses were about, "in a madhouse, screaming might be the *normal* means of communication."

The running youth began to loom larger during the second week. He'd been seen by another witness who'd been walking his dog, a witness who claimed the youth looked as though he was being chased. Described as a teenager, five feet seven inches tall, with dark collar-length hair, this one may or may not have been the original runner. The police realized there could've been several young men running home on such a cold night.

They began tracing anyone at all who'd been in the general area that night. A teenager had been seen getting off the bus outside a pub in Narborough. The driver could say only that he'd picked the boy up in Huncote on the 599 bus at 6:38 P.M. Nevertheless, he was hunted for days.

At 7:10 P.M. on the night of the murder, a young woman had boarded a bus from the bus shelter on Forest Road. The driver wasn't certain, but thought she'd got off at Foxhunter Roundabout near Enderby. She was sought for weeks as a possible witness to verify the report on the young couple allegedly seen at the bus stop. Then there were two women, one in her early twenties, who'd boarded the 600 bus to Leicester. They too were hunted in vain.

After buses the murder squad started on scooters. A teenager had been seen pushing a motor scooter past the psychiatric hospital just after 8:30 P.M. on November 21st. He'd worn a long green parka, but he didn't seem to have a crash helmet so he might have been pushing it to a garage. They sought out *all* youths with motor scooters, whether or not the scooters functioned.

The *Leicester Mercury* was of great help, printing virtually whatever the police wished. And of course, each printing brought hundreds more calls, all assigned a priority, all given to various teams.

Nearly every day either Baker or Coutts was interviewed by reporters, and made public pleas: "I urge people to cast their minds back to the evening of November twenty-first. . . ."

Before the second week was finished, the murder squad had checked out hundreds of reports. One of them concerned two teenage boys who'd bought a copy of the *Mercury* from a newsagent's shop in Narborough on the afternoon the body was discovered.

"The lads studied the paper very intently," police were told. "They should be investigated."

They were.

Still another young man was sighted *twice* on the evening of the murder, once on Forest Road and another time walking toward the hospital. His priority was raised.

And at 7:30 P.M., just after Lynda was last seen alive, a man carrying a guitar case had been seen sitting across from the chemist's shop in Jubilee Crescent. He was added to the list.

By the third week, the police were making even more appeals to the public through the newspapers and television. They particularly wanted the running youth.

"Perhaps some young man arrived home out of breath after ten o'clock that night," Supt. Baker suggested to reporters, "and ran straight upstairs to avoid his parents."

There was a "crying youth." He'd been spotted near the murder scene five days *after* the crime, sitting at curbside opposite The Black Pad. A couple driving by had seen him and immediately telephoned the incident room. He was a fair-haired lad, about seventeen years of age, wearing a bomber jacket. A motorcycle was propped up by him. The crying youth was not found. Boys his age wouldn't come in to admit to such an unmanly display.

The newspaper pleas started paying off. A guitar player called the incident room to see if he was the one they were trying to trace. More running youths were reported, including a new one who'd run under the M1 motorway bridge. And soon the murder squad began hearing about runners and punks from as far away as Birmingham. They were inundated with punks and runners. Given tips on punks who sounded like Johnny Rotten, they'd more often than not track down a youngster with dyed sideburns and an ear loop, who was just going through a phase.

On December 15th it was announced that lights would be installed on The Black Pad at a cost of £5,500, and on the same day an unnamed relative of Lynda Mann made a personal appeal to readers of the *Leicester Mercury*. The headline read: PLEASE HELP TRACE THIS MANIAC.

During that third week in December the police were offered a "Teddy boy." A new witness had spotted a couple standing on a corner of Leicester Road in Narborough at about 8:20 P.M. on the night of the murder. When the driver slowed to allow them to cross, the youth said something to the driver, no doubt something cheeky, because the driver described the youth as being similar in appearance to the youthful rebels of an earlier generation.

Then there was yet another young man who'd bought a copy of the *Mercury* and "made suspicious inquiries as to whether or not there were details of the murder in it."

It was notable that almost all reported suspects—hundreds of them—were youths. Village people obviously saw the murderer of Lynda Mann as someone quite close to her age, and in fact, so did the murder squad commanders. All of them—the punks, Teddy boys, runners, criers, weepers, readers—all of them were teenagers.

That December a large number of officers volunteered to keep the incident room open during the holidays, even on Christmas Day.

Supt. David Baker took the occasion to say to the media, "Christmas is a time of year when people start reflecting. Lynda's family will certainly be looking back, and also the person responsible, and his family. We would urge anyone who notices anything manifestly different about family members in the Narborough area to come forward and inform us."

He then went on to suggest for the first time that Lynda's killer had been known to her: "They were probably acquaintances, and perhaps what started off as a kiss and cuddle developed into something that got out of hand, resulting in Lynda's death. But only the person responsible can tell us what actually happened."

The police were thus openly offering extenuating circumstances to the killer or to anyone who might be shielding him. There were no takers.

---

That Christmas, Kath Eastwood had some presents to give out, presents that had been bought by Lynda. Ever the enterprising, resourceful, and self-sufficient girl, Lynda had taken money she'd saved from babysitting and bought the presents well in advance of the holiday. Kath gave them out on Christmas Day.

By now the Eastwoods desperately sought what most victims of cruel and terrible crimes want: retribution *and* revenge. Eddie and Kath were always honest enough to admit the latter.

Eddie told reporters who rang him that Christmas Eve, "We live each day hour by hour, minute by minute. I just hope the man who killed our Lynda is suffering as much as we are. I just hope he's thinking about the damage he has done to our family while he celebrates *his* Christmas."

Kath also gave a statement: "This man has got to be punished. And I hope anyone who knows him will think twice about harboring him. We just do not have an existence anymore and he is to blame."

---

The huge Edwardian brick buildings of the psychiatric hospital, with their gray slate roofs and eccentric campaniles, looked ugly to some, especially that massive brick chimney towering over the countryside. But Derek Pearce said, "I found the old place quite handsome, except it looked very *eerie* coming in at night."

The eeriness was no doubt heightened by thoughts of the poor wretches confined in those buildings. Perhaps even him, the one they hunted.

More than one detective was to describe driving into the hospital grounds on dark brooding nights, thinking of him and wondering if he was peering out a window. Watching and laughing. If you were tired from overwork, if you'd had a couple of pints, it wasn't impossible to fancy you'd heard a soft demented chuckle in the darkness, from just across the cricket pitch.

# 7
## Plea

By the first week police were desperate enough to ask the *Mercury* to print more pleas for witnesses to come forth. The first request involved two men and a young woman who'd been observed in a coffee shop on Horsefair Street in Leicester. One of the men had been reading an inside page of the *Mercury* when he abruptly folded up the paper. After the young woman asked him what was in the news, he hushed her by saying he'd tell her when they got outside. The witness who'd observed the incident told police the page the man had been reading was "probably" page 13, which carried a story on the Lynda Mann inquiry. The young man wore a gold earring in his right ear. There were *lots* of earring wearers reported.

Still another published plea asked for information leading to a "mystery man" who'd scribbled the name of Lynda Mann in a telephone book in a local kiosk the day her body was found. That particular clue fizzled when the mystery man telephoned police admitting he'd jotted down the name while ringing a village friend to ask if he knew the victim's family.

By the end of January the police were publicly releasing well-worked information on a youth who'd entered a wool shop in Narborough the day after the murder for a new pair of trousers because his were torn. The shopkeeper's suspicions were belatedly aroused because of a published report that police were looking for a beer-bellied young fellow with a tear in the left leg of his jeans, who'd been seen coming from The Black Pad at 8:35 P.M. on the night of the murder.

Now even anonymous calls were prompting large newspaper stories. In early February a young woman rang the incident room at midnight to inform police breathlessly that she knew someone resembling the spiky-haired punk with orange hair cropped like a bunch of geraniums. She had seen the artist's impression in the newspaper and was sure he frequented a public house in Enderby. The police interviewed everyone in and around the public house for a week, but the only geraniums they saw were potted.

By February the murder squad was still nearly one hundred officers strong. They'd taken three thousand statements and followed up some four thousand lines of inquiry. Virtually every young man going through a punk phase in Leicestershire and the surrounding counties had been interviewed. In the beginning they had thought that if they ever found one who was five-ten, slender, who wore boots with laces, and a leather jacket, and a belt with a bronze buckle, and had amazing orange hair, it would *have* to be their man. But they had found *lots* of them, all with amazing hair, many with laced boots and leather jackets. None was the punk seen with the girl thought to have been Lynda Mann, the punk who had caused the motorist to brake sharply.

He was, in the words of a team member, "as elusive as the flippin Loch Ness Monster." And the running youth had worn out several teams. They said they'd interviewed more runners in early 1984 than had the British Olympic coaches.

Supt. Ian Coutts was still convinced the spiky-haired youth was their man and that the girl seen at the bus

shelter had to have been Lynda Mann. Derek Pearce
and many of the others weren't so sure, but everyone
believed the killer must be a local man in order to have
known about The Black Pad and the gate leading into
the wooded copse beside that tarmac path.

As far as Coutts was concerned, Lynda had probably
been friendly with her killer because she wasn't the sort
of girl to talk to a stranger, and would've fought for her
life if suddenly ambushed. They searched endlessly for
a "secret boyfriend," one not known even to her best
friends. Someone with whom she might have taken a
stroll, along The Black Pad.

The Police Mobile Reserve is a unit of uniformed
police officers drawn upon to supplement the divisions,
a pool of men for any job. The PMR did the house-to-
house *pro forma*, and took statements from anyone not
alibied.

On January 22nd, Police Constable Neil Bunney of
the PMR had on his list a semi-detached house in
Littlethorpe, part of a new housing estate in a street
called Haybarn Close. The owner of the house was a
twenty-five-year-old baker named Colin Pitchfork who'd
recently moved into the house with his wife and baby
from Barclay Street in Leicester, about five miles down
Narborough Road.

Pitchfork's young wife, Carole, answered the door,
admitted PC Bunney, and called upstairs to her husband. Everyone in the three villages knew that the
police were doing house-to-house inquiries, so the constable didn't have to explain much.

The baker didn't come down for several minutes.

"I had to compose meself," he later said.

What really had the baker worried was that he'd been

up in the attic putting down floorboards he'd stolen
from a construction site, telling his wife he'd bought
them at a bargain sale. He thought for sure he was
about to be nicked for the theft, and it wasn't the only
thing he'd stolen; he'd also pinched a cabinet unit he
thought might fit nicely in the kitchen. He wasn't ordinarily a thief, but the opportunity had presented itself
and he wasn't one to pass up an opportunity. Upon
talking to the policeman, the baker was extremely relieved to learn that the officer had no knowledge whatsoever of the stolen property.

In that Colin Pitchfork hadn't lived in Littlethorpe at
the time of the murder he would've been relegated to a
low-priority classification, except that he had been discovered to have had a prior indecency record. It seemed
that he was a convicted flasher, and had been from a
very early age, so he was on the list. Later, after a
computer match-up, he ended up on three indexes: the
indecency list, the Carlton Hayes outpatient list because he'd been referred to therapy by the court for
one of the flashing offenses, and the house-to-house
resident list.

Pitchfork's classification in the incident room was as
an unalibied "code four." Code one meant that the
suspect couldn't have done it because he was dead, in
prison, or his whereabouts had absolutely been proved.
Code two meant he'd been alibied by a friend or colleague. Code three was a wife's alibi, which was never
very reliable. The inquiry team could end up with a
man who was part code two, part code three and part
code four, if, for example, he'd seen his wife early in
the evening, gone out with friends and then walked
home alone.

They were looking for someone who was unalibied
between seven and midnight. According to both Pitch-

fork and his wife, he had driven her to a night class at the community college early that evening while their baby slept in the backseat in a carrycot. The baker then had gone home to his former residence and sat with the baby until his wife was finished with her class. Technically, he *was* unalibied from 6:45 P.M. until 9:15 P.M., but he would've had to leave his baby unattended in order to go out and murder.

Psychologists maintain that flashers are a relatively harmless lot, and Pitchfork had no history of violence. Furthermore, he'd moved to Littlethorpe in December, one month *after* the murder of Lynda Mann. Not having been a villager at the time of the murder, he probably wouldn't have known about The Black Pad and the gate into the copse.

He was not given a high priority, but his diabolical surname caused a joke or two in the incident room. After all, a Pitchfork would have to be guilty of *some* sort of villainy.

---

Derek Pearce's father had always wanted him to become a doctor. The older man had been a strict parent: ex-army, railway worker, traffic warden. He'd ended his working career as the curator of the regimental museum in Leicester. Pearce's mother was, as Derek called her, "a mum's mum." Pearce had a brother one year older and two younger sisters, but four weren't enough for Mrs. Pearce. She became a foster mother, and the house was literally crawling with babies. She'd take any and all kinds, with or without birth defects.

When Pearce was nineteen he'd joined the Leicestershire Constabulary on a whim. As a result, his father hardly spoke to him for a year. When his older brother changed his college program from biochemistry to medicine, Derek Pearce was relieved. He hoped that with a budding doctor in the house his father would relent.

A lifelong problem with his inner ear made it difficult for Pearce to walk a straight line. In marching drills at the police training center they'd put him in the middle and let him bang into the shoulders of his colleagues, but he was good at other aspects of police training.

During his first year in the field he won the Harris Cup given to the probationer of the year, and his picture appeared in the *Leicester Mercury*. That dose of celebrity mollified the old man. Nobody in the family had ever been photographed for the newspaper. His dad was very proud.

By his second anniversary Pearce was transferred into CID, and was promoted to sergeant three years later, with an exam score among the top two hundred in all of England. He made inspector six years after that. Everyone said he was a "flier."

Then the flight got diverted. Promotions beyond the rank of inspector are based not on written examinations but on scores given by panels of senior officers, as well as on written recommendations from immediate superiors. Derek Pearce's annual reports were very good, but troublesome words popped up occasionally, words like "arrogant" and "intolerant."

Pearce summed up his management role by saying, "For me life should be nicking villains and being a cop. If theirs wasn't, they were working for the wrong DI."

Nobody doubted his ability to do police work, and Pearce looked after his people by defending them against all outsiders. He was generous in a pub and was good to them when they needed an afternoon off, but he could be ruthless with any subordinate who treated police work "as a job rather than a way of life." If they worked hard and made only honest mistakes he'd administer a verbal "bollicking" that usually went no further. But his bollicking was about as subtle as a wrecking ball.

Stimulus wasn't often needed during the Lynda Mann inquiry. Members of the murder squad maintained they never lost confidence that they'd detect their killer, convinced he had to be a villager. And though it was nearly impossible to match Pearce's intensity for crime detection, he often tried to ignite his subordinates with his unabated energy. Even after long, fruitless, frustrating days Pearce always looked forward to tomorrow.

"What about *this* idea?" Pearce would say, eyes dilating as he seemed to rise up from his chair, hovering, *levitating*.

He'd often toss them an idea the others hadn't tried. But if he didn't like his subordinates' ideas, Pearce was canny enough not to discourage them, "unless they were too one-off," as he put it. Pearce believed his job was to keep his detectives enthusiastic, hopeful, excited.

"Where is he? We know he's right here, don't we? What would *you* like to do? What do *you* think? Never mind what the gaffer thinks, I'm asking *you*. Our man's close by us, isn't he? I can *feel* him. Where is he? Where is he, then?"

They knew he was manipulating them, but strangely enough, it kept them enthused despite themselves.

"Nobody said you had to love him," one of his men said later. "You just felt like throttling him sometimes when he started shouting at you, but he could organize. He could always see the big picture immediately. He had a computer for a memory and could sort things out, even if he did talk to you like a bleedin foxhound."

"Where is he?" Pearce would say. "He's right here, isn't he? Come on, let's find him! And, my lads, let's not forget our happy little home."

He could be right there in the hospital itself, Pearce

often reminded them. Where they might have more gibbering loonies than a Labour party picnic. More perverts than the House of Lords.

The irrepressible Derek Pearce seldom talked about his ex-wife even with close friends, even if he'd been mixing his drinks. She'd also been a police officer, a few years younger than he, very attractive and with a personality every bit as strong as his. It was a disastrous mix. Everyone who knew him said the torch Pearce carried could've ignited glaciers.

After she walked out on him she'd needed a down payment for a cottage. Pearce wrote her a check for £1,100, and when the purchase somehow fell through, he wrote another one for £1,100 to go with the first. The solicitor handling Pearce's divorce rang him at that point and said, "I'd like a letter from you stating that you go against all of my advice. I *need* it for my professional reputation."

When she left him she wouldn't tell Pearce where she was living, but if Derek Pearce was anything, he was a good detective. He searched probable neighborhoods and spotted a vase in a window, a vase she loved. When he found her she was sick in bed with the flu, and had no one to take care of her.

"Why not let me come round and look after you till you're better?" he suggested.

"Maybe," she said.

"And when you're well you might come back home . . . for a *visit*. To say hello to the dog."

"Maybe," she said.

She got some looking after, all right. When he was through scrubbing, that cottage was clean enough to impress Joan Crawford. But his former spouse *didn't* go back home to say hello to the dog. She went to live in Hong Kong with another man, who was a former colleague of Derek Pearce. The torch still flickered.

---

Kath and Eddie Eastwood had been trying for weeks to get the coroner to release Lynda's body for burial. They were repeatedly told, "We must maintain control of the remains until all forensic work is completed."

Eddie said it just went to show how the authorities treat poor people, but Kath said stoically, "I *suppose* they know best."

Perhaps, but in the length of time Kath was denied her daughter's body, they could have taken apart the two-hundred-piece skeleton of Lynda Mann bone by bone. They could have dissected every organ, grouped and subgrouped five quarts of blood drop by drop. The hair could have been catalogued strand for strand, and clothing fibers subjected to more scrutiny than the Shroud of Turin. Whatever they needed or *thought* they needed, the coroner's people maintained custody of the body of Lynda Mann for more than ten weeks.

Finally, on the 2nd of February, Kath was allowed to bury her daughter in the cemetery by All Saints Church—a few minutes from where she'd lived, a few steps from where she'd died. More than a hundred people, including Supt. Ian Coutts, attended the funeral. Several other detectives observed, and made a video of the mourners, looking for what, they weren't certain.

"We got the best stone we could afford," Kath Eastwood said. "We expected it to cost two or three hundred pounds, but it cost nearly eight hundred."

Eddie said, "It were over one thousand quid all together, the funeral. I called Social Security for help and they says they can only spend twenty! 'She wasn't stillborn!' I told them. 'You can't bury a fetus for twenty quid!'

"We go to the grave regular," he said. "It seems daft to talk to a grave, but people do. It helps."

"It's somewhere to go," Kath Eastwood said. "It brings solace to my mum. She likes to visit Lynda's grave."

The carved inscription on the heart-shaped stone read:

LYNDA ROSE MARIE MANN
*Taken 21st November 1983*
*Aged 15 years*
*We didn't have time to say goodbye,*
*but you're only a thought away.*

Kath kept all of Lynda's clothes. Some people told her to get rid of them, but she couldn't. Eventually Eddie put them up in the attic.

"I kept having a dream," Kath said. "I dreamed of Lynda fighting. Of being dragged down."

She wished she could dream of other things, perhaps a dream of Lynda bringing a cup of coffee to her bedside on Christmas morning. That was the kind of memory she wanted to relive in dreams, but the recurring dream was always the same. Of Lynda being dragged down by a looming shadow without a face.

# 8
# Visions

By mid-February the murder squad had distributed a thousand posters of an artist's impression of the spiky-haired youth, and had put together a twenty-minute video about the murder which they showed at local schools and shopping centers, and even at a disco where youths gathered who might've known Lynda. The video described several of the promising leads and witnesses they had yet to locate, primarily the spiky-haired youth.

By late February they had a brand-new one: "the somber girl." This lead was phoned in to the incident room by a witness who'd spotted a young couple strolling by Copt Oak Road on the night of the murder. It looked to the witness as though they'd been arguing, because the girl was in "a somber mood." The man was six feet tall, slim, and wore an "unfashionable coat," according to the caller. The girl, of course, was thought to be Lynda Mann.

Suddenly they had a new sighting of Lynda reported, this one in Leicester city center where she'd been reportedly seen with a punk who had three-inch spiky bleached hair. It was known that Lynda used to go into Leicester every Saturday, so the police treated this one seriously.

The various leads were driving the inquiry's man-hours into the thousands. It was perhaps with a note of desperation that after showing the video in a disco at Croft, the *Mercury* announced that murder squad spokesmen were saying, "We are closer." But they were not.

They took their show on the road; the video was seen at more local shopping areas and schools, more discos in the surrounding areas, and even by shoppers in Leicester city center.

In March, Supt. Coutts was telling journalists that the girl seen at the bus stop on the night of the murder *was* Lynda Mann. After examining thousands of leads, Coutts had to believe in something. Because the punk never came forward, he would remain the strongest lead. Ian Coutts had it fixed in his mind that he was one of the two people seen at the bus shelter, and the other *had* to have been Lynda.

"She wasn't the kind o' lass to go wi'out a struggle," the Scotsman repeated to the end. "He must o' been someone she knew."

During the Caroline Hogg murder inquiry Derek Pearce had learned for the first time how to look at pedophiles, discovering through that exhaustive and futile investigation that there were far more sexual deviates living in the villages than he'd ever thought possible. A sexual offense was reported every day, and since everyone knew that the number of reported sex crimes never reflects the true extent of the problem, he always wondered how many occurred. How many in Narborough, or Littlethorpe, or Enderby? How many had their man himself committed before he'd killed Lynda Mann?

Blood testing was done on many of the most promising suspects. The best they could get out of the forensics laboratory was "He could've done it" or "He might've done it." A subtlety that escaped Pearce.

Sometimes they'd be told he could *not* have done it,

because they couldn't find the PGM 1 + factor. Yet
even if police sent in a sample *not* from the PGM 1 +,
A-secretor group, they were told that another 40 percent of the blood group couldn't be ruled *out*. Science
was vague, ambiguous, mysterious. The police believed
they'd never get an answer from scientists with all their
"probably's" and "possibly's" and "maybe's." They blood-
tested some of the huge number of workers brought in
to build the new housing estate by The Black Pad,
many of whom lived nomadic lives in tinker caravans. It
required enormous investigative time to verify the alibis of these itinerant workers.

There were many theories and arguments about
whether or not the chest bruises, one darker than the
other, could have come from the killer's knees while
Lynda Mann was being strangled, and many debates as
to whether or not the bruise on her chin was caused by
a blow of sufficient force to knock her out. Those in
favor of the knockout theory had to deal with the biting
of her tongue. That was covered by saying it had happened when she was coming around, the instant he
applied the ligature. All of this was a subtle way of
questioning the opinion of their commanders that Lynda
must have known her killer.

One night in The Rosings they were brainstorming
while Derek Pearce was in his frenetic bird-dog mode.
"Come on, lads, what do you think? How do you fall?
Let's come *up* with something!"

So someone did: a tall, rangy detective constable who
parted his red hair in the middle and let it hang down
lank on both sides. He looked like one of those earnest
schoolmasters from the old prewar films they ran late
nights on the telly. Or, as one detective put it, using

the ultimate epithet of cops the world over: "like a social worker."

What the DC came up with was: "Has anyone considered that a woman might have done it?"

It may be that Derek Pearce's brown eyes pulsed and swelled behind his eyeglasses. Maybe that repertory company beard of his started to twitch, or maybe he simply began his trick of levitation. Whatever, someone sensed a moment was forthcoming and switched on the tape recorder.

Pearce waited until all eyes were on him, until not a few cops leered like expectant hyenas. Then he said, "Yes, there might be a *lot* of women in these villages who prowl about at night with a plastic bag full of come concealed in their purses." He paused to let them consider that before continuing: "Mind you, not just common *ordinary* come. Oh dear, no. But bags full of PGM one-plus, A-secretor come!" They say he was a foot off his chair for the denouement: "And then they take their bloody missiles of PGM one-plus bloody come, and like some demon bloody bowler, *they hurl it at the crotches of the victims they bloody well strangle before they stroll home for tea and bloody biscuits!*"

The brainstorming was over for the evening. The tape machine was switched off. Everyone gathered up belongings and put things in drawers. The DC who dared to look like a social worker decided to call it a day. They all left Pearce to be spotted by Heathrow traffic controllers as he hovered between the campaniles and the towering chimney.

———

The murder squad had been decreasing in size as the clues petered out. From 150 officers, they'd dropped to

50, then 30, and by April they were down to 16, and suddenly they were 8.

The Caroline Hogg inquiry had entailed a lot of work and ended in disappointment, but they'd always believed her murder had been done in Scotland and the body dumped on them. Lynda Mann was different. She *belonged* to them.

"The answer is right *here*," Coutts always said, and Pearce agreed.

"I never came to work," said Pearce, "without thinking that *today* we find him."

But by Easter it was essentially over. Coutts went to Derek Pearce and said, "We've got to close it down. They won't let us go on." Then he added, "I havena' come in second many times in ma life." He choked back whatever else he wanted to tell his inspector.

At the last assembly in the cricket pavilion where he had so often addressed them in freezing weather with steam blowing from his lips, Ian Coutts thanked his officers, told them how grand they had been and what a privilege it was to have worked with them. And when he began to express his personal regret that they would not be allowed to continue, the tears welled and he hastily concluded and left the room.

Some of them said it was difficult to believe that the Scotsman could cry. They said he was the archetypal soldier Scot, so hard and tough. But Ian Coutts had been born near Glasgow, a city known for tough, friendly, *emotional* people.

"If ever a man deserved a result from hard work, it was Jock Coutts," Pearce said. "He put all he had into the enquiry on Lynda Mann."

When it was over, DSgt. Mick Mason returned to Wigston Police Station after personally seeing to it that the files were transferred to a nominal Lynda Mann incident room set up in the communications headquarters of the Leicestershire Constabulary.

By summer, the eight officers had dwindled to only a pair: Mick Mason and one other. They had other duties, but still took occasional calls from someone who'd spotted a spiky-haired youth, a running youth, a crying youth, a husband, a boyfriend, a brother-in-law, or a neighbor with a fiendish glint who'd ogled a matron suntanning in the backyard of a cottage in Enderby.

By the end of the inquiry they'd given about 150 blood tests to potential suspects. The results were more than disappointing.

In August it ended altogether. When it was over they had a list of about thirty suspects who were "possibles." There were no "probables."

---

Before the incident room was closed and the murder squad was disbanded and returned to divisions, there was one contact made with a villager that later proved noteworthy. A lad, fourteen years of age, was reported by several villagers as being a "bit of a nuisance." He was a curious youth, usually seen riding around the village on a bike with cowhorn handlebars and derailleur gears. Sometimes he prowled the streets and parks and footpaths. He liked to jump out at women and girls, to frighten them.

"He's always mucking about" was how it was put by more than a few villagers.

Mick Mason interviewed the boy. He lived with his family in Narborough, by the Foxhunter Roundabout. He used to be interested in CB radio but as he was getting older he longed for a motorbike.

He didn't seem overly bright, was quiet spoken and

didn't give the detective any trouble. He answered perfunctory questions and Mason was satisfied.

Mick Mason had a mental picture of Lynda Mann through contact with Kath and the family. Mason had made regular visits to the Eastwoods, and when they'd come to the incident room, Kath would always walk up to Mick and give him a hug, because, as Pearce put it, "he's that sort of bloke."

The village youth was a big lad, a bit thick, a bit odd, but Mason was absolutely sure he was not the one.

"Lynda would've been able to sort out a fourteen-year-old," he told Derek Pearce. "Any fourteen-year-old."

Mick Mason clearly admired and respected Lynda Mann. He had come to know a girl in death he'd never known in life.

---

The Eastwood family turned to a medium. She was fortyish, frail, rather timid. The medium came to their house and sat in Lynda's bedroom. Kath and Eddie stayed on the stairs and listened to eerie noises within. The medium held one of Lynda's necklaces and made terrible choking sounds, but the Eastwoods had to remain outside and not interfere. When the woman emerged she said, "He was a big strapping man. He came up from behind."

She refused to take money for reporting her "vision," and said she'd return and try again some other time. Before she left she said, "The afterlife is on a different plane. We all live on different planes, those of us in this world and the others. This world is hell."

Kath didn't doubt *that* now, not for a moment.

"And where's Lynda?" her mother asked.

"She's in the other plane. It's like being in hospital

there, the equivalent of hospital. She'll continue living there much as she was here."

The woman was obviously trying to put them at rest, but the family was upset for days. A big strapping man coming up behind Lynda. The choking, strangling sounds. It was like Kath's recurring dream!

"If he's not caught within one year, he'll do it again," the medium warned.

---

The aftermath of murder produces many casualties. Susan Mann had, like Eddie Eastwood, been center stage during the inquiry into the death of her younger sister: Lynda the pretty one, the bright one, the popular one, the one they doted on.

The police inspector responsible for the night patrol of Narborough began receiving reports of a shapely girl in a miniskirt flagging down patrol units as well as CID cars. When he talked to his officers he learned that the girl was Susan Mann who was offering "clues," none of them worthwhile.

"Leave her alone" was the word he quickly passed.

A sergeant who'd become a friend of the family spotted Susan one night and decided to have a chat with the girl, who had never *openly* grieved for her murdered sister.

"Look here, Sue," he said, "we've been ordered not to talk to you anymore. It's just not *on*. You can't be waving down police cars."

"*Nobody* talks to me anymore," she said. "I even talk to Lynda, in me own mind, but nobody talks to *me*. Nobody cares!"

"Well, of course we do," the sergeant said.

And suddenly, Susan Mann grieved. Perhaps for Lynda, perhaps for her mother or for herself. In any case, the lonely girl cried. She wanted comfort and a shoulder. She reached for him.

"Steady on!" he said. "This won't do! I'm a married man and a bloody policeman!"

So they all wept eventually, *all* the victims of The Black Pad killer. They all wanted someone to hold them and tell them it was all right, even though they knew it wasn't and would never be again.

# 9
# Discovery

Derek Pearce had, in the course of his police career, read books dealing with forensic medicine, like Dr. Simpson's. But like the majority of policemen, he was skeptical that science would ever do more than occasionally augment "old-fashioned bobbying," police work being more of an art than a science to its practitioners.

"I'd always heard about startling scientific discoveries," he said, "but I'd never been startled. A fingerprint in blood was about as startling as it ever got for me on a murder enquiry. I always said I'd like to see one of those scientists startle me someday."

He was about to get his wish. A few miles away from the village of Narborough, a thirty-four-year-old scientist at Leicester University was, in the autumn of 1984, about to stumble upon a discovery which would indeed startle Derek Pearce, as well as every other man and woman in the Leicestershire Constabulary.

The discovery occurred on a "fringe project" taking place in the laboratory of geneticist Alec Jeffreys, whose main project involved a study to determine how genes evolve. He was working with those genes which express specifically in human muscle. He was most interested in the repetitive sequence he'd found in the human myoglobin gene.

His little fringe project, progressing to its final stages, was of minor academic interest. Then it suddenly became a bit more intriguing.

Several years earlier, genetic engineering techniques had been developed for looking directly at genes and deoxyribonucleic acid—their genetic material—known as DNA. Geneticists had become interested in the possibility of examining the genetic differences between people, the most fundamental aspect of all.

The DNA molecules that govern heredity were troublesome and elusive because most genetic material varies quite a lot from one person to another. The challenge, as Jeffreys saw it, was to identify those regions of genetic material that displayed the most variation from person to person, and then to construct a means of highlighting these regions with a radioactive probe. Simply put, Jeffreys was trying to develop much better genetic markers than had yet been found, specific markers for mapping human genes.

There was a bit of the philosopher in the geneticist. Two years before he began work on his fringe project, he'd been quoted regarding genetic engineering being done at Leicester University: "Too many people still throw up their hands in horror about the ethics or morality involved without really knowing what genetic engineering is really about. What we've really got is a new toy with endless possibilities and it's just amazing what could be done. New medical compounds could be made, crops and farm animals could be manipulated to provide better yields, and many diseases could be virtually eradicated."

Alec Jeffreys then tried to allay the fears of people with Frankenstein fantasies. "It is the scientists themselves who first thought that experiments could be dangerous, but experience has shown that very little can go wrong. There's no way scientists could contemplate using humans as guinea pigs, even in the unlikely event that their governments would let them. I have this nightmare vision of people choosing the characters of their children in the same way that they choose a car or a washing machine. But I think society will always make sure that science does not interfere too much with nature."

Jeffreys's twenty-seven-year-old lab assistant, Victoria Wilson, had been with him for eight years by September of 1984.

"The years were just a blur," she later said of those earlier times. "They just sped by, but Alec seems to somehow remember everything we've done, just as it happened day by day."

The biology scientists are "interesting types," according to Vickie Wilson. But even the most interesting among them must have been startled the first time they saw the bearded geneticist with a roll-your-own dangling from his lips. He looked anachronistic, like a pot-smoking academician from the sixties, but it wasn't cannabis, it was Golden Virginia tobacco. He let the cigarettes go dead in his fingers when energetically involved in conversation, and never smoked tailor-mades unless going somewhere special.

He was seldom seen without his bulky turtleneck. That "polo neck jumper" became a Jeffreys trademark.

The geneticist's staff consisted of a research assistant, two technicians, usually a couple of Ph.D. students, and a postdoc or two, all of whom enjoyed his style.

"Alec's always excited," another technician said of her boss. "He's *such* an enthusiastic scientist. His personality inspires the rest of us."

Whenever he got excited, Vickie Wilson figured he was on to something, and he had been *very* excited one day in September of 1984.

"During the course of that research Alec was poorly," she remembered. "He had glandular fever, and I had to ring him up quite often as things were proceeding. To tell him it was getting *quite* interesting."

Jeffreys's process for mapping those human genes entailed taking DNA molecules extracted from a sample of blood cells, and cutting or "chopping" them into unequal bits by adding enzymes to them. The fragments were dropped into an agarose gel where an electric field caused the larger fragments to separate from the smaller ones. The DNA fragment pattern was then transferred to a nylon membrane by a technique called southern blotting, literally drawn up by capillary force when the blotting paper was placed on the membrane.

Jeffreys's team then added radioactively labeled pieces of the DNA to act as "probes" that would stick to the hypervariable regions they fitted. The membrane was X-rayed to disclose the radioactive pattern, the darker bands appearing where the probes had adhered.

The distribution of these bands would be unique, person to person, and so they would be looking at a DNA image that would be individually specific.

That was the theory of how it was *supposed* to work.

On a Monday morning in September the X-ray film was developed and, in Jeffreys's words, "We were just stunned!"

Within minutes of getting the film out of the developing tank they could read it! Furthermore, they had expected to see one or two major bands on the film, but instead they had a whole series of gray and black bands, resembling the bar codes used to mark grocery items. And Dr. Alec Jeffreys knew that he was looking at huge numbers of genetic markers that showed both an astonishing level of variability and an amazing degree of individual specificity.

Jeffreys's wife, Susan, was a senior computer officer at Leicester University, and it was she who realized the possibilities of his discovery. She immediately predicted the practicality of what he'd called his "lucky string of circumstances." By that evening she was making long lists of the applications of his technology, the first being to suggest that immigration disputes could now be easily settled, and in Britain there were many. It would now be a simple matter to determine whether or not a person seeking entry into the country was entitled to do so, based upon an alleged blood relationship to a British subject.

It wasn't long before Jeffreys and his team theorized that the system might be used on animals, with very substantial implications in determining pedigree, in artificial insemination, in ascertaining that endangered species didn't inbreed accidentally.

It could be used in bone marrow transplants of leukemia victims to determine whether grafts had taken or not. It would be easy to determine if newborn twins were fraternal or identical, since the only people on the face of the planet with identical DNA maps would be identical twins.

And it didn't take them very long to see that the technology could have important uses in forensic analysis. A name for the technology had to be chosen. The bold and logical choice was "genetic fingerprinting."

On November 19th the *Mercury* printed a column headlined: STORIES TO JOG YOUR MEMORY.

The newspaper article offered a news summary from Monday, November 21, 1983, the day of Lynda Mann's death. There was a story about multimillionaire Soraya Khashoggi, née Sandra Daley of Leicester, who'd made a bid in court to have her ex-husband jailed. That was one locals would remember. Another old story dealt with the Leicester rugby hero who'd skippered his team, the Tigers, to a victory over Twickenham.

And for several days the paper ran a series of articles summarizing the massive work done on the inquiry, ending once again with an appeal for help. The police admitted to "relying very heavily" on the local press.

The Leicestershire Constabulary had initiated a new poster campaign and set up a mobile inquiry unit to receive any fresh leads. The police asked people to "search their consciences to remember what members of their families and friends were doing on Monday night, November 21, 1983."

Another Lynda Mann poster was displayed throughout the villages. The poster heading said, "Let's not forget Lynda Mann, murdered a year ago in Narborough." The poster showed a photo of Lynda's face superimposed over a model posed in a donkey jacket. The poster asked the public to help with the inquiries and ended with the promise that all information would be treated "in the strictest of confidence."

Derek Pearce reported to journalists that the renewed campaign had resulted in thirty telephone calls, but he belatedly added: "Some lines of enquiry are *not* new."

On a dreary autumn day, one year after the murder of Lynda Mann, Kath and Eddie Eastwood visited a second seer. This one was a woman they'd heard about from friends in Leicester. The Eastwoods were not trying primarily to find comfort in otherworld contact, but rather looking for murder clues, to alleviate rage and despair.

The woman was very vague and offered little solace. She reported a vision involving the initial "T." and the name "Gerard." She could see a pretty dark-haired girl, as though in a mist. But the Eastwoods realized that she could well have read about the murder and could have seen photographs of Lynda. The medium also warned that if the killer was not caught very soon he'd kill again, but the police told them that anyone with a rudimentary understanding of such crimes might have made a similar prediction.

The "T." meant nothing to Kath and Eddie. The "Gerard" they thought could refer to the notice for Gerard Motors over the railway bridge on Narborough Road. They wondered if perhaps the murderer lived near that bridge. The police, however, were not enthusiastic about seers and mediums.

Sgt. Mick Mason still kept in touch, and they received some calls from Inspector Mick Thomas, the other young DI who had worked with Pearce on the Lynda Mann inquiry. Thomas would ring them from time to time offering reassurance that the police would never give up, that they were working on every new lead and reworking old ones.

He'd even call when one of the new leads *didn't* pan out. The calls that admitted failure went a long way in convincing them they still might hope for retribution. If he'd just called with optimistic information, they'd have chalked it up to public relations, and felt more hopeless than ever.

They got a bit of comfort when they learned that a published report claiming Lynda had been seen in a disco on the night of her death had been investigated thoroughly and discounted. She'd been to a disco in Croft with her friend Karen on November 18th, but that was the extent of it. They thought it *had* to have been a villager who'd murdered her, not some stranger from a disco.

*        *        *

During that month, reporters interviewing the East-woods printed a story that "financial difficulties and health problems" had beset the family during the pre-ceding twelve months. Eddie told of hopes being dashed that his work would permit him to take the family to some other part of England. He was now resigned to staying there in Narborough.

"Kath can never look in the direction of The Black Pad when we're driving by," he said.

Kath's mother, the sixty-four-year-old grandmother of Lynda Mann, told reporters that she visited her granddaughter's grave once every fortnight to replenish the flowers. She said, "As a Christian I do feel sorry for anyone who has done this. It may have been something that got out of hand. But I feel bitter too. It has crip-pled us. He *must* be brought to justice."

Kath said publicly that even former friends tended to avoid her and the family. She didn't blame them and understood their reasons.

"They just don't know what to say to you," Kath explained. "They feel that they *must* say something and they don't know what."

But she longed for her old companions to return.

Just after the first anniversary of the death of Lynda Mann, a hospital worker from Carlton Hayes Hospital found a tiny cross with a poppy attached to it, there in the ground beside The Black Pad, on the spot where Lynda had been murdered. Nobody knew what to make of it. The Eastwoods thought it might be a sick joke. Others thought it might be a gesture of remorse on the part of the killer. Poppy Day in England fell two weeks before the anniversary of the murder so the police thought it was just a simple gesture by some child, but who could say?

# 10
# Breakthrough

One of the first experiments Alec Jeffreys conducted using genetic fingerprinting was on a family group to see if the pattern of inheritance was as simple as he expected it to be. From that test he saw clearly that half the bands and stripes on the X-ray film were from the mother, and the rest from the natural father. The patterns were inherited in a sensible fashion. It was *thrilling*.

Determining constancy from tissue to tissue within the individual followed next. His team took both blood and semen and found that the genetic map was constant irrespective of the kind of cells from which the material had come. To discover how sensitive the system was, they tested small quantities of blood and semen. It was *rather* sensitive: a drop of blood was enough, or a tiny amount of semen.

But there was the question of whether DNA was stable enough to survive in degraded forensic material. Jeffreys had conversations with forensic scientists at the Home Office who had access to three-year-old blood and semen stains. They tested these and it worked again.

Then they began testing the system on a wide range of animals and fish. Again it worked, and as they improved and refined their system the resolution and clarity of the X ray got even better. It only remained for the excited geneticist to write up his discovery for publication in the scientific press. He did the writing, but held up publication until he had his patents; there were highly profitable implications to his discovery.

Jeffreys didn't speak publicly about genetic finger-printing until November, 1984, one year after the death of Lynda Mann. He discussed it then at two meetings in London: one with the Lister Institute of Genetic Medicine, and the other with the Mammalian Biochemical Genetics Workshop. These satisfied the scientific disclosure requirement for his patent application, an application that listed Jeffreys as the inventor and the patent rights as vested in the Lister Institute of Preventive Medicine, of which he was a research fellow. Alec Jeffreys wanted any commercialization to benefit a British company, so Lister selected Imperial Chemical Industries as the sole licensee for any and all commercial exploitation.

In March 1985, when the Leicester geneticist published, he estimated the chances of two people having the same DNA fingerprint, two brothers for instance, as literally none. "You would have to look for one part in a million million million million million before you would find one pair with the same genetic fingerprint," Jeffreys said, "and with a world population of only five billion it can be categorically said that a genetic fingerprint is individually specific and that any pattern, excepting identical twins, does not belong to anyone on the face of this planet who ever has been or ever will be."

It was a dramatic claim and brought an immediate practical test. Shortly after publication, Jeffreys was called upon to enter an immigration case, a complicated one involving a boy who was living in Africa with his father, but who'd been born in Britain of Ghanaian parents. The boy wanted to return to Britain and live with a woman he claimed was his mother, but immigration officials believed the woman was his aunt and he'd been denied British residency.

Jeffreys had somehow to match the bands in the child's genetic fingerprint with those of a father not present. And he had a mother who wasn't all that sure about the boy's paternity in the first place.

The geneticist decided to take the undisputed children of the woman and match their genetic fingerprints with the mother's to "reconstruct," with some measure of certainty, an absent father. When he compared the pattern of the boy with those of his siblings, the reasonable conclusion was that the man who'd fathered him was the same man who'd fathered the rest of the children.

Journalists loved that one. They wrote articles suggesting that some of the huge disputes on *Dallas* and *Dynasty* could easily be resolved by genetic fingerprinting. It was said that Dr. Alec Jeffreys had done a disservice to crime writers the world over, whose stories often center around doubtful identity and uncertain parentage.

Later in 1985 Jeffreys published again, after another system was developed in California, called the polymerase chain reaction. The California technology was even more sensitive than Jeffreys's system, which could get down only to a single hair root. But the California system, which had actually produced a genetic fingerprint from forty sperm heads, didn't have as high a level of individual discrimination. For forensic analysis, the Jeffreys system needed larger amounts of quality genetic material, but its end product was highly discriminatory. The California system could work on more degraded genetic material and so had its own place in forensic science. The bases of the two systems were very different, but they complemented each other.

Jeffreys wasn't afraid to test his system in a high-profile forensics matter, but the attention he was receiving in 1985 was basically confined to the scientific community. He got a professorship from Leicester University as a result of his discovery, was awarded a string of medals and prizes, and was admitted as a fellow of the Royal Society.

When the Home Office publicly accepted evidence provided by Jeffreys as convincing enough to use in deciding immigration cases, Alec Jeffreys was quoted as saying, "This makes me very hopeful that it will become a recognized method."

His comments were printed in the *Leicester Mercury*, two years after the murder of Lynda Mann. The last paragraph in the Jeffreys article said, "The new technique could mean a *breakthrough* in many areas, including the identification of criminals from a small sample of blood at the scene of the crime."

---

A month prior to that second anniversary of murder, while Alec Jeffreys was preparing a second paper for publication in a scientific journal, a sixteen-year-old trainee hairdresser said good night to her teenage boyfriend late one evening on a street corner in Wigston, a few miles east of Narborough. The boyfriend kissed her and ambled off, disappearing around a corner. The hairdresser turned toward home, walking down Blaby Road.

*She looked fourteen or fifteen to me. A pretty little brunet. I was having a wander. Just driving down Carlton Drive when the opportunity presented itself. I followed behind her, but Blaby is a main road. Don't touch her on a main road! Mustn't touch her on a main road!*

The hairdresser walked into Kirkdale Road and then

turned right toward the footbridge over the railway, toward Kenilworth Road. There was no lighting and she moved more quickly through the darkness on the footbridge.

*It had to be right. The car had to be parked in a line with the girl between me and the car. That's the way it had to be. I parked the car in an alley at the back of some shops.*

Coming off the bridge the hairdresser saw something that made her break stride. She saw a shadow figure pacing up and down on the right side of the pavement. It seemed a bit peculiar. The shadow figure seemed to be waiting impatiently. When she got close to him she saw he was wearing a dark-blue nylon hooded jacket with white tassels. She was glad there were houses just ahead. The hairdresser ignored him and tried to walk past.

His arm shot out and hooked around her neck! She started to scream. She felt the blade of a screwdriver against the left side of her neck. A powerful hand seized her mouth. He whispered, "Shut up screaming or I'll *kill* you."

The blade gouged, the hand smothered. Like rabbit and fox, she was jerked, shaken, dragged along the pavement—dragged backward between the houses, back toward a row of garages. When they reached the opening he carried her inside, and pushed her forward in the darkness until her face pressed against the damp brick wall.

He released her mouth tentatively and said softly, "It'll be all over in a bit."

"Why me?" she pleaded. "Why me?"

"You're the only girl around, aren't you?" he said reasonably.

"They'll all be coming out of the pubs soon," she sobbed. "There'll be *more* coming along! Why *me?*"

He began moving his free hand all over her. He fingered the zip on her trousers. She whimpered and pushed his hand away.

"You do it then," he whispered. "*You* do it."

She instinctively pulled back but stumbled in the darkness. She found herself sitting on the concrete floor, weeping.

"Take my money!" she cried. "Take *anything* but leave me alone! I promise I won't tell anybody!"

"Well," he said, dispassionately, "you'll have to suck it now, won't you?"

She started to get up but he shoved her down, letting her feel the blade while he unzipped.

She didn't really believe he was going to do it. It all seemed so impossible she refused to believe it. He stood over her. He shoved it in her mouth. It was flaccid. He withdrew and masturbated.

*She never shouted. She never screamed. I took her to the floor very gently. I told her she was stupid to be walking home alone at that time of night. And what was her boyfriend letting her walk around at that time for? And where the bleedin hell were her parents? And why didn't they come to fetch her? They're nowt but silly twats, I told her.*

He put it back in her mouth. She didn't move. She wanted to clench her teeth but was too frightened.

*"Let this be a lesson," I told her. "You never walk the streets at night. You might not be enjoying this but I ain't hurting you. I could've knocked you off dead easy."*

Suddenly he withdrew from her mouth. He turned away for a second. She didn't know if he ejaculated. If he did, it was when he turned away, or perhaps he *didn't* at all.

*She knew what I said was right. She agreed with it. I came, but I can't remember where I did it. She pulled away from me.*

He said, "Don't say anything to anybody or I'll come back and find you."

The hairdresser sat there sobbing after he'd gone. She believed he might be there waiting to test her, waiting in the darkness. She managed to stand and move. She crept from the garage. He was gone.

Her parents were already upset that she'd been seeing the boyfriend and staying out late, so she didn't tell them what had happened. But while at work in Leicester the next day she had to tell someone. After blurting it out to another hairdresser, she felt better. She also told the manager of the salon, and the manager called the police.

————

There were some pathetic local headlines in 1985. The *Mercury* ran a story with a headline that read:

MURDER STARTED ROAD INTO DEBT

Edward Eastwood got into debt because he was unable to work after his stepdaughter, Lynda Mann, was murdered, Leicester magistrates were told.

Eastwood (43) of The Coppice, Narborough, admitted five offences of obtaining credit while an undischarged bankrupt. He was remanded on bail until August 5 for social inquiry report.

After the first bankruptcy he obtained credit from five different companies without disclosing he was bankrupt, but he had made repayment on all the loans. Solicitor Mr. Walter Berry said Eastwood got into debt because of the very traumatic situation which affected the whole family after the murder in November 1983.

A doctor's letter, which he handed to the magistrates, told a 'terrible tale of woe,' he said.

Before the murder, Eastwood was working as a quality control manager and was putting in 90 hours a week to pay off his loan debts. But afterwards, the court was told, he was unable to work and lost the job.

A month later a story headline said:

MURDER TRAUMA LED TO OFFENCES

The stepfather of murdered teenager Lynda Mann was ordered to do 150 hours' community service by Leicester magistrates after he pleaded guilty to five offences of obtaining credit while being an undischarged bankrupt.

Prosecuting, Mr. John Davis said Eastwood had obtained credit to buy a greenhouse and furniture, have car repairs done, as well as obtaining a loan to pay off debts, and he was subsequently declared bankrupt again.

Magistrates read probation reports and ordered him to perform 150 hours' service to the community during the next 12 months.

As to his misfortune, Eddie Eastwood said, "English law and the people who administer it have no finesse. No respect for people like me."

It may well be that the travails of Eddie Eastwood would have taken place regardless of the events on The Black Pad, but there are subtle changes that take place among survivors, as all families of murder victims know. Murder, particularly the murder of children, often produces a complicated, even insidious emotional aftermath.

———

On the second anniversary of Lynda Mann's death, an unknown person once again placed a small cross there in the wooded copse beside The Black Pad.

# 11
# The Kitchen Porter

Like the neighboring village of Narborough, Enderby has an old stone church that's outlasted everything else, the parish church of St. John Baptist. One of two churches in Enderby, it's just across the road from a Church of England school, long since closed down. Like the one in Narborough, the little churchyard of St. John Baptist is full of old headstones, but in modern times the vicars needed more burial ground and so a cemetery was consecrated behind the church.

The seven Enderby pubs, a lot for a village this size, served the needs of hundreds of quarry workers when quarrying was the area's main industry. Many granite walls in Enderby were cut from those quarries and still exist in the old part of the village.

Rows of terraced brick homes, formerly belonging to quarry workers, stand as good examples of substantial no-frills Victorian housing. As many as twenty-five are sometimes blocked together in a row, differing only in the color of paint on doors, rain gutters and window casings.

With seven boozers in the village, the young people have their own. Inside theirs, the decibel level from recorded music, teenage guzzlers and fruit machine gamblers could terminate pregnancy.

But a short walk toward the old part of the village, where the roofs are slate instead of tile, leads to quiet cozy pubs with decent kitchens for a bit of beer-soaked steak or "boozy beef." The New Inn still has a thatched roof, and The Dog and Gun was established in 1650. The pavement in that part of the village is three feet wide, and the two-inch "curbs" hardly qualify.

It seems certain that Enderby will change from a village to a town long before Narborough does. It isn't that the village has so many more people than the combined total of Narborough and Littlethorpe, but that the residents aren't as steadfast in retaining the character of a village, having allowed more shops and businesses. And it isn't just that Enderby has more of the quarriers' terraced homes and fewer Georgian houses and Tudor cottages. It has to do with attitudes.

The police say that the young people in Enderby are more "anti" toward the cops, whereas the youngsters in Narborough are more "pro"—friendlier—Narborough being more middle class. Yet some newcomers to the villages say that the adults in Enderby are more welcoming, being more working class then their neighbors in Narborough.

The residential streets and lanes of both villages are a mixture of architecture and economic circumstance. The teenagers of both villages attend school at Lutterworth, a town with many lovely Georgian buildings, some six miles away.

For their part the police deal with class distinctions by defining homes on their burglary computer as "council, private, or very high class" for the upmarket and crusty.

———

It was difficult to say to which village he belonged, living there by the Foxhunter Roundabout. His mailing address was Narborough, but Enderby was just as accessible to a highly mobile teenager. During the spring of 1986 the boy was no longer riding about village streets on his bicycle with the cowhorn handlebars. He was seventeen years old by then and had gotten his first motorbike.

He was six feet tall, built large in the hips and thighs. He still never bothered to comb his scruffy hair, and didn't change his jeans very often. His clothing was spotted with grease and oil even more than before, and his fingers were encrusted with grime now that he was tinkering with engines. With physical maturity his brow was a bit more overhung, but he'd probably long have the boneless expression of a child. For sure, he smiled a child's smile, a *secret* smile, like a giggle suppressed. He was still known as a quiet-spoken loner.

But one of the villagers, the local locksmith, saw with his own eyes that the "nuisance" hadn't outgrown his bothersome ways. The locksmith was browsing with his small daughter in the video shop a few doors from his home. The tiny shop did double duty, for it also served as the Narborough Taxi Company. A young woman handled the videos for hire and dispatched taxis.

The locksmith noticed the boy enter the shop and go over to the video collection. A teenage girl happened to bend over to pick out a cassette, and the boy moved behind her, running his hand up between her legs.

Everyone in the shop was shocked except the boy. He just showed his secret little smile as if to say, "Anything wrong?"

The locksmith, formerly a British Army judo instructor, strode up to the boy and said, "I'll bang your head in if I *ever* see you do that again!"

He didn't respond. He looked at the man and smiled.

"He just *stared*," the locksmith later said. "As though he didn't quite understand. I would never let my daughter anywhere near *that* lad."

*   *   *

Despite his love for motorbikes, despite being nearly full grown, the seventeen-year-old seemed to prefer the company of young children. There was a family of six living just down Narborough Road, a family he liked to visit after school. The oldest boy was two years younger than he. The youngest girl was nine years of age. Though he ran errands for her parents and sometimes gave her rides she didn't like him very much.

According to the child, "He looked and smelled like a fish, so I called him Fishface."

One afternoon she was playing outside when he arrived on his motorbike to visit her older brother. She had a hairbrush in her hand and after they did some name calling back and forth she walked up and smacked his motorbike with the brush. He took the brush away and shoved the bristles against her nose.

"Who's a fishface?" he taunted, then slapped her across the face.

She ran and told her father, who ordered the boy to go home and warned him never again to put his hands on the children.

But the following Saturday, he returned. The little girl called him Fishface once again and he chased her down on foot and tackled her. She bit him.

When she talked about it later, she said, "He pulled me knickers down and put his finger in me money box. It hurt! He were wearing black motorbike gloves."

A fourteen-year-old girl, who'd seen him pursuing the child, tracked them and found him with his hands inside the child's pants.

"Get off her or I'll kill you!" she said, picking up a stick and striking at him.

The boy called them both a few names, including "slag," ran back to his motorbike and sped for home.

Six days later he came back when the little girl's parents weren't home. This time he entered the house and said he was going to wait for her older brother to return. The little girl's eight-year-old brother was there, and both children told him to get out. The nine-year-old girl again called him Fishface.

He hit the brother who ran upstairs crying. When he was alone with the little girl she punched him, and dashing through the house to get away, she fell and bumped her head. While she was lying on the floor crying, he knelt down and pulled at her underpants.

"He poked and hurt me inside me privates again," she later said.

The seventeen-year-old had stripped her underpants down to her knees when the family dog growled and leaped on his back. With one child crying upstairs, one screaming downstairs, an Alsatian snarling and barking, he retreated to his motorbike.

The nine-year-old girl was afraid to tell her mother about the incident, but described it to her thirteen-year-old sister. The sister wanted to inform their parents, but the younger child cried and made her sister promise not to tell.

Then one afternoon in May of 1986, the child was down the road with a teenage girl who fancied another motorcyclist. The three of them were off on a side road in a secluded spot near a housing estate. The older girl and the other cyclist were sitting on a motorbike kissing when Fishface drove up on his motorbike.

When the seventeen-year-old offered to trade the little girl a bite of his chocolate flake for a kiss, she hopped on his bike and kissed him. After all, sometimes he was nice to her. He'd even given her a necklace for her ninth birthday. But then he began fondling her. He stopped when she began shouting for the others to make him get his hand out of her pants.

It appeared that he was still having a lot of dangerous problems with *them*, the ones he called slags, dogs, whores, and bitches. No matter how young they were, they just didn't like him.

"He's not very bright," his mother later said of him. "He's a bit down in his age group, education-wise. But he's *all right*."

He was *not* all right. But he was employed. He'd gotten a job at Carlton Hayes Hospital as kitchen porter, and they referred to him more grandly as "catering assistant."

It seemed to amuse him. When asked about his job, he'd show his secret little smile and say, "I'm a kitchen porter. I hand out food in a lunatic asylum. I work in a loony bin!"

Before the end of the school term in 1986, the police issued a new appeal through the *Mercury* for village residents to come forward with clues in their relentless hunt for the murderer of Lynda Mann.

The article said:

> Even though it was such a long time ago, police still hope that someone, casting their mind back to that Monday night, can come up with the killer.

And once again, a police spokesman reiterated the unswerving opinion:

> Poilce are convinced she had a pre-arranged engagement with someone on the night of her death, and that she was the girl spotted with a man at the bus stop in Forest Road, Narborough,

between 8:05 P.M. and 8:30 P.M. on the night of
the murder.

The police were soon to be dissuaded of that opinion.

———————

Eddie Eastwood and his family had not recovered
from their economic reversals and talked continuously
of getting away from the village, making a fresh start in
some other part of England. Eddie still had trouble
finding and keeping suitable employment and still complained of his arthritis, which had gotten steadily worse.
During an unproductive summer, in the waning days of
July, Eddie was offered a day job by a local farmer. The
job entailed mowing a field of young seed hay, which
was to be harvested later and used for horse feed. The
field was between the M1 motorway and a footpath that
provided a shortcut between the Narborough and
Enderby village centers. It was a pleasant walk along
that footpath. A gate opened from it onto the field that
Eddie mowed.

The footpath was called Ten Pound Lane, or sometimes Green Lane, because it was so overgrown, so
lush and lovely in summer.

Within three days of Eddie Eastwood's mowing, that
field would be swarming with police, and Ten Pound
Lane would become more feared than The Black Pad.

# 12
## Ten Pound Lane

In the Enderby home of Robin and Barbara Ashworth was what some thought to be the sweetest family photo they'd ever seen: The Ashworths and their two children, Dawn and Andrew, were standing in a row, their arms linked, each with a genuine smile. The handsome family had posed in front of the bay window of their terraced home in Mill Lane. There were sixty panes of glass in that Georgian bay window, devilish to clean, but a nice architectural touch. The Ashworths had a spacious four-bedroom house and a large bungalow on a third of an acre, with gardens out back.

On July 31, 1986, Robin Ashworth was forty years old. An engineer for British Gas, he was introspective, placid, boyish looking. Not given to pique or temper, he was the sort who, schooled in science, believed in being reasonable and logical with his children. Dawn, who was just fifteen, and her lanky thirteen-year-old brother, Andrew, had a blend of both parents' coloring. Neither had their father's tousled, charcoal-brown hair, nor their mother's fair hair and blue eyes. The children's hair was more of a coffee brown. Dark-eyed Andrew was a quiet boy, polite, rosy-cheeked like his mother. Dawn's eyes were blue-hazel, often described as bright and expressive, suited to her effervescent personality. Neither of the Ashworth children had ever given much bother to their parents.

As Barbara Ashworth put it, "We thought we'd had

every blessing. Robin had a sister and I had a brother, so when we had Dawn followed by Andrew, it was ideal. Perfect. Just right."

Having been an older sister herself, Barbara said, "I often thought things were unfair when I was a girl so I always saw Dawn's side of a disagreement. I'd try to get down to her level and view things the way she did. If there was a row, it was between Dawn and me to work out. In my era we didn't discuss a lot of things with our mothers, but in this day and age you can. I'd go upstairs and cry on Dawn's shoulder, and she'd do the same. We'd just clear the air."

They were a family who believed in talking about problems. Robin in the deliberate, reasonable way an engineer might present a proposal at the gas works, Barbara with a big-sister chat, followed by a few tears and a hug. They were a rather good mix, Robin and Barbara Ashworth.

Some three months earlier Barbara had changed jobs and was now being kept busier working customer liaison at Next, a purveyor of upmarket fashion and one of the fastest-growing businesses in the UK. With headquarters in Enderby, Next had introduced a new shopping concept to Britain, the idea that a woman could go into a large store, have her hair done, buy coordinated wallpaper and furniture, and then shop for clothes, all in one place. When Barbara had applied for the job it was part-time and she had a lot of time to tend to her large garden. As the job took on more hours, becoming full-time on Monday, Tuesday and Wednesday, it meant that Dawn had to do more housework, provoking a few disputes between mother and daughter.

When the school term had ended in June, Robin and Barbara had been disappointed with Dawn's grades. Robin's way to handle it was to take Dawn to task,

present clearly and concisely why he judged her perform-
ance substandard, and then leave it. He wasn't the
type to complain to his daughter about school grades.

"I never expected Dawn to pursue an academic career," he later said. "She was gifted at drawing. More
of an artistic type. She could do a picture without even
being taught about perspective and things like that. She
could really *see*."

Actually, Robin was not all that certain about many of
the roads he'd taken. He didn't know if he should've
gone to university instead of getting his advanced education in night school. Would it have meant a better
position? He wasn't a man to let off steam. Was *that* the
proper way? He pondered such matters.

Robin was, as he put it, "the kind who worried about
small things. I used to feel nervous and apprehensive
about jobs and things at work. Enough to make whoever I was with feel the tension."

Never having been quite sure about decisions he'd
made or postponed in his own life, he couldn't fret too
much about his daughter's underachieving in academics, particularly since she was generally sensible and
rather mature, perhaps more comfortable with herself
than he had ever been with himself. A reasonable man
could hardly demand much more than a child who
could really *see*.

During the first week in July, Robin had taken the
entire family to Tall Trees Caravan Park in Norfolk
where they'd stayed in a friend's caravan. Dawn had
been excited about another holiday trip to Hunstanton
that was coming up. She had a part-time job that summer working at the newsagent's shop in Enderby, and
she was spending her money on all sorts of fashion
magazines and clothes. The Ashworths had regularly
received good reports from Dawn's employer at the

shop. She was said to be a reliable, likable girl who seemed to know everyone in the village.

In high summer, Dawn began going out nearly every evening to the home of her friends in Narborough. Robin and Barbara didn't like her being out so much even on summer's light nights, but Dawn had agreed always to be home by 9:30 sharp. Occasionally, if there was something special afoot, Dawn would ring her dad and ask for a lift.

On the morning of July 31, Robin woke his daughter to go to her job. She was a bit cross and grumpy, complaining that he should have woken her earlier.

Dawn did her job that day as usual, and nothing extraordinary occurred in the shop. An employee later said that two girls came in whom Dawn was temporarily "on the outs with," and Dawn had a few mildly catty things to say about them, but it was just adolescent bickering.

The Ashworths lived only a few minutes from the newsagent's shop, and after receiving her wages at 3:30 P.M. Dawn walked home. She told her mother she was going to have tea with her friends Sue and Sharon, in Narborough. Barbara told Dawn to be home by 7:00 P.M. because she and Robin were going to a birthday party for a friend's little boy.

That made Dawn decide to add some sweets to the present she'd already bought for the child, so she returned to the newsagent's shop and bought a fifty-pence box of Smarties to include with the other small gift. She also bought a pale-pink lipstick.

She had ten pounds with her when she left the shop at 4:00 P.M., heading for Narborough. She was wearing a white polo neck pullover, covered by a multicolored loose-fitting blouse. She wore a midcalf white flaring skirt, white canvas pumps, and carried her blue denim jacket.

The most direct route to the homes of her two girlfriends in Narborough was by way of the footpath. Dawn had frequently been warned about the village footpaths by her parents.

"She was aware that there was a killer about," her father said. "Dawn was years ahead of herself and never queried important rules. She was well acquainted with the Lynda Mann case."

Two of the village girls saw Dawn that afternoon, walking past Brockington bowls and tennis courts, heading toward Ten Pound Lane. There'd been a Midlands summer shower that afternoon, the kind where one minute there's brilliant sun and the clatter of birds, and next there's rain slamming on the roof, followed almost immediately by silver storm light.

A teenage boy also saw Dawn that afternoon as she neared the verdant footpath. He said that Dawn was a cute and bubbly girl, but he hesitated to speak because he didn't know her well enough. "Her hair was sticking up on top as if she had gel on it," he later said.

Dawn reached the fork and had a choice. She could walk left on the path over the motorway and then parallel with the motorway to King Edward Avenue—or to the right, toward Ten Pound Lane. Since the Lynda Mann murder her father had repeatedly told her always to walk over the motorway. But it was broad daylight and she was growing up, and, in her mother's words, "blossoming and changing, day to day." She chose the shortcut and walked down Ten Pound Lane, which was sodden from the shower and smelled of rotting leaves.

Dawn emerged from Ten Pound Lane, crossed King Edward Avenue and cut through the hedge to Carlton

Avenue. She called at the house of Sharon Clarke, knocked and was met by Sharon's mother.

"Is Sharon there?" she asked.

"No, she's just gone with Sue," Mrs. Clarke said. "Perhaps you'd like to try Sue's house."

"I will," Dawn said. "Bye!"

A few minutes later she knocked at the door of Sue Allsop's house.

When the door opened Dawn said, "Hi. Are Sue and Sharon there?"

"No, I'm sorry, dear, they're not here," Mrs. Allsop said. "Probably gone to the village. Why don't you go look for them?"

Dawn Ashworth decided not to look for her friends in Narborough village. A neighbor of the Allsops standing in her kitchen saw Dawn heading back toward the motorway.

A passing motorist later said he sighted Dawn Ashworth at 4:40 P.M. crossing King Edward Avenue, walking toward the farm gate, about to enter Ten Pound Lane.

———

*I was riding the motorbike when I saw her cross the road. She walked through the gate there to the lower footpath. I parked the motorbike a short way from the main road and I put me hat on the handleclip. When I walked through that gate a gut feeling was saying, No no no no no! But the other side of me was saying, Just flash her. You've got a footpath. You've got all the time in the world. Even if she runs off screaming no one will ever see you. No one will ever know! Who's going to know?*

———

Ten Pound Lane was perhaps the loveliest of the village footpaths. By the lower road, it was entered through a wooden farm gate. At that point, the path had been covered partway with black tarmac about two feet wide, but soon it ran out and you walked on a grassy dirt track.

On the motorway side were fields of hay dotted with poppies, sprinkled with a bit of heather. And on the other side was a mini-golf course, and then farmland bordering the Carlton Hayes Hospital, protected by a five-foot fence. A profusion of nettles and brambles, along with birch, elm, and tangled hawthorn bushes, on both sides of the path created a veritable tunnel of green where the path got narrow, where it snaked toward the psychiatric hospital and away from it.

Villagers were sometimes solicited in newspaper ads to buy pyracantha and berberis. The Royal Society for the Protection of Birds wished to provide "life-saving berries for your hungry garden visitors." There were pyracantha and berberis both, tangled in the impassable wall of blackthorn that bordered Ten Pound Lane.

It was an ideal place to walk a dog. The whole path from King Edward Avenue to Brockington School took only fifteen or twenty minutes to cover, walking briskly. And on the upper side, toward Enderby village center, there were lights and a playing field for soccer. But before you emerged at that point, the net of foliage would brush your face as you walked. And you had to pass through the dark places where only a few spangles of dappled sunlight filtered down onto the footpath.

Some of the trunks in the thickets were a foot in diameter where the path narrowed to one foot, where it was most overgrown and studded with jutting rocks, where you passed through a narrow green tunnel. The sky over your head could actually disappear for a moment in that lush green tunnel on Ten Pound Lane.

\* \* \*

Robin Ashworth left work at 4:40 P.M., and a short while after he got home, the phone rang. It was Sue Allsop asking for Dawn, explaining that she hadn't been home when Dawn dropped by. Robin told Sue he was sorry but Dawn hadn't come home yet.

After Sue rang off he decided to take the dog for a stroll. Sultan was an English setter he'd bought Barbara as a birthday present, but as Sultan grew older he needed a man's voice to keep him in control, so he'd become Robin's dog. Robin and Sultan walked the footpath that led to Blaby Road.

When Robin returned, he changed clothes for the party, but when Dawn wasn't home by the 7:00 P.M. deadline, her parents were worried enough that Barbara went alone just to deliver the birthday presents. Barbara returned anxiously at 7:30 P.M., but Dawn still had not returned.

Barbara then drove to Sue Allsop's house and learned from Sue's mother that Dawn hadn't been seen since 4:30.

Barbara Ashworth wasn't quite frantic, not yet. Not until she searched the village streets and found both Sharon and Sue on a seat opposite the newsagent's shop in Narborough. The girls had no idea where Dawn could be.

Then Robin and Barbara began to search in earnest with the help of friends. Robin even walked the footpath bridge by the motorway, by Ten Pound Lane. Three or four joggers passed him as he walked. When he covered The Black Pad later that night it was hard to stay in control, hard not to think of the *other* girl, even though in high summer the light nights had brought to middle England a beautiful silver sky, with a pale wash of crimson cloud.

When the search proved futile, Robin and Barbara

wanted to call the police at once, but realized they'd be asked what time Dawn was *supposed* to be in. They knew they'd be advised that Dawn had probably forgot they were going out, and would be home by the regular time: 9:30 P.M.

They waited until 9:30, but Dawn did *not* come home. They rang the police at 9:40.

Just as Eddie Eastwood had done on a freezing autumn night in 1983, Robin had searched all the logical places and tried to retrace Dawn's probable route from her friend's home in Narborough. Just as Eddie Eastwood had done, Robin Ashworth walked the footpaths. And just as Eddie Eastwood had done, Robin walked within a very short distance of his daughter, lying by the field Eddie had just mown, nearly halfway down Ten Pound Lane.

---

The next day, Friday, August 1st, found a large number of local policemen and tracker dogs searching the Narborough area for Dawn Ashworth. The significance of senior detectives being present at the footpath was lost on no one. A footpath by the mental hospital. A sensible fifteen-year-old girl who'd been happy at home. *Déjà vu*. The fields between Narborough and Enderby and the tree-lined footpaths were searched with negative results.

Robin and Barbara Ashworth spent Friday at home with two police inspectors who gave them what Robin called "a fair old grilling." The police didn't seem to accept anything they were told at face value but went off and checked each fact concerning Dawn Ashworth and all of her friends.

The house, garden shed and bungalow were searched,

even up to the eaves in the roof. They didn't ask for an article of Dawn's clothing for their tracking dogs, which seemed to be trained only to look for disturbed undergrowth. A policewoman examined every page of the Leicester telephone directory to see if Dawn had marked or underscored anything of significance. As with any missing fifteen-year-old, it was generally surmised that she might have run off, especially given no evidence of foul play.

"No," Barbara Ashworth told them. "She's *not* run off. When she was five minutes late I knew something was wrong."

"She knew we were going out to see friends last night," Robin Ashworth told the detectives. "If she didn't want to walk home, she'd have rung up. She knew I'd come round and pick her up. Dawn never hesitated to ask me for a lift."

One of the detectives said, "She may've just gotten a bit angry with you—about the extra chores you said she's been given lately."

"I tell you there's something *wrong!*" Barbara Ashworth repeated endlessly that day.

Still, the policemen looked as though they'd heard it all many times before, as indeed they had. A detective looked at a school photo of Dawn taken two years earlier and asked, "Is this a good likeness?"

"She hates that one," her mother said, "but it's a good enough likeness."

The policeman smiled reassuringly to help hold them together and said, "Oh, when she gets back she'll be so annoyed at you for giving us this photograph, won't she?"

Robin's fair old grilling had to do not only with where he'd been the previous day, which was meticulously followed up by the detectives, but where he'd been on the night of Lynda Mann's murder. Robin Ashworth got a bit of what Eddie Eastwood had gotten nearly three years earlier, but the police were sensible enough not to probe too far in that direction.

During the prior year, Dawn was known to have had two or three casual boyfriends, but was never known to have had any sexual experience. About the most serious peccadillo anyone reported to the police was that Dawn had been seen smoking a cigarette the prior week out in front of a boy's house, and she reportedly had kissed a boy at a pajama party. She was *not* the sort of girl to have suddenly run off.

"Dawn wears a brace on her upper teeth, does she?" a policeman asked Barbara Ashworth.

"She's worn it for a year," Barbara said. "She'll be having the brace taken off very soon. On August thirteenth."

"She'll be happy about that," said the policeman.

That afternoon the newspaper headline read:

HUGE HUNT FOR MISSING SCHOOLGIRL

Senior detectives and uniformed police with tracker dogs have joined a huge search of the Narborough area for a 15-year-old girl who disappeared last night not far from the spot where another schoolgirl was found murdered three years ago.

Dawn Amanda Ashworth, of Mill Lane, Enderby, has not been seen since she visited friends in Narborough yesterday afternoon. She left their house on Carlton Avenue, Narborough, at 4:30 P.M. and disappeared.

That night at 10:55 P.M., as a result of the newspaper story, the Ashworths got their first phone call. Barbara answered it. There was no voice on the line.

"Is it you, Dawn?" she cried. Then she shouted, "Robin, get to the other terminal!"

Robin picked up the second phone and said, "If it's you, Dawn, we want you home! Please, Dawn! If we've done anything to upset you please come home and we'll talk about it!"

The thoughts of Barbara Ashworth were veering crazily. Maybe she's had a bump on the head! Maybe she's got amnesia! Maybe . . .

Then Robin said, "If you're holding our daughter, please just put her down anywhere! Unharmed! I *beg* you!"

The person on the telephone may have tired of it. The line went dead.

Exactly fourteen hours later, at 12:55 P.M., the phone rang again. The person still refused to speak.

"I beg you, please!" Barbara Ashworth sobbed. "Don't hurt our daughter!"

"Just let us know she's all right!" Robin pleaded. "That's all we ask!" It was a reasonable request, and he was a reasonable man. He may have expected a reasonable response, but got none. No response at all.

The police arrived and decided that the person might be someone doing shift work, ringing them up when he was going to or coming from work. The police stayed and took a similar call for them at 4:30 P.M. that afternoon. The detective told the Ashworths that those things often happened in such cases.

The newspaper headlines on Saturday, the 2nd of August, were growing more ominous:

DAWN: FEARS GROW FOR HER SAFETY

More than 60 police, some with tracker dogs, are now involved in the investigation. They are concen-

trating on house-to-house enquiries while surrounding fields are being searched, and Dawn's friends
interviewed.

The spot where she disappeared is five fields
away from the lonely Black Pad footpath where
the body of another 15-year-old girl, Lynda Mann,
was found nearly three years ago. Her murderer
has not been caught.

And there was a plea from RobinAshworth printed by
the *Mercury* that day:

Dawn's distraught father, Mr. Robin Ashworth, a
scientific officer for British Gas, said, 'If anybody is
holding her, at least let us know she is safe. She
would never have gone anywhere by herself. She
always respects what we say.'

Robin ended his plea by saying, "She will be panic-
stricken by now, from being away for so long!"

That morning, while searching the freshly mown fields
between the M1 motorway and Ten Pound Lane, a
police sergeant found a blue denim jacket by the footbridge that went over the motorway. There was a lipstick and cigarette packet in the pocket. The entire area
was sealed off immediately by wide bands of orange
tape, and by uniformed constables.

# 13
# Square One

Before noon that Saturday morning, several police officers encircled a clump of blackthorn bushes in a field beside Ten Pound Lane. A bank of freshly cut hay, broken nettles, tree branches and other foliage had been heaped atop the blackthorn, all but concealing the body of Dawn Ashworth. They could see only the fingertips of one hand.

Like that of Lynda Mann, the body was naked from the waist down except for underpants still on her right ankle, and the white pumps still on her feet. She was on her left side with her knees pulled toward her chest. Her bra was pushed up to expose her small breasts and there was a smear of dried blood extending from her vagina across her left thigh. She wore only one earring, a silver three-quarter flattened hoop.

There were numerous injuries to the body, many of them postmortem, from insect bites and from being dragged through stinging nettles. The body showed generalized rigor and a temperature of 64 degrees.

Nature had tried very quickly to claim Dawn Ashworth. Relentless crawling insects had savaged her so badly that at first detectives thought she'd been brutally beaten. Implacable flying insects had deposited eggs in every orifice of what had been a vibrant human being. Like Lynda Mann's, her eyes were heavy-lidded as though she'd been grieving.

One arm was outstretched in front, the one wearing a

wristwatch. Robin and Barbara had often told friends of the joy they'd gotten in choosing that watch for a Christmas present, the last Christmas their child had seen. The watch displayed the correct time.

On the Friday night that Dawn Ashworth was missing, there was a party at the home of a Leicestershire policewoman. Derek Pearce had been one of the revelers in attendance. But he was thinking about the missing fifteen-year-old. It was like Lynda Mann all over again, a failure he'd never gotten over. That evening another inspector said to him, "We'll find her dead somewhere."

As the party progressed, Derek Pearce found himself talking to "a lovely young lady with big eyes." He couldn't take his gaze from those big eyes. In fact, they may have made him a bit giddy. Or he may have been listing to starboard. He leaned back against a wall.

No wall. It was a door that opened into the bathroom. There was a step *down*. Pearce later described his move as a "pirouette." Others described it as a Benny Hill pratfall. A human skull collided with a toilet bowl leaving a visible crack in each.

True to form, Pearce resisted the partygoers who wanted to rush him to hospital.

"No way I'm going!" he said to Sgt. Gwynne Chambers, who was several years older than Pearce, and a friend on and off duty.

Chambers showed Pearce the blood and said, "You're *going*."

The emergency staff of the Leicester Royal Infirmary worked on him from 1:30 to 5:45 A.M., finally burning the vessels to stop the bleeding.

He woke up Saturday morning. The first thing he saw was the ward sister bringing a bottle to his bedside.

"What's that?" he asked.

"For urination."

"I've got some pride," Pearce informed her. "I'm *not* peeing in a bottle."

And so it started. He tried to get up, she pushed him down. One word led to another and he leaped out of bed. He was *weak*.

She smiled and said, "*Now* will you get back in bed and behave like an adult?"

He reached up and found his hair matted with dried blood.

"I'm having a shower," he said, later admitting that he felt like a tire puncture in the rain—out of control and skidding.

"Promise you won't get your head wet!" she demanded.

Never a great line-walker due to his lifelong inner-ear problem, he now walked like a racetrack pickpocket, bumping into anything in his way and fumbling for handholds. He managed to shower but failed to tuck the curtain inside the stall. The sister found bloody water washing down the floor of the corridor.

She stormed into the room and shouted, "You promised faithfully you wouldn't wash your hair!"

"Where's my clothes?"

"We've taken your clothes."

"I want my clothes."

She stalked out again and he staggered across the ward but had to return to bed.

An hour later he opened his eyes and was shocked to see Chief Supt. David Baker and Supt. Tony Painter sitting beside him.

"You look awful," Baker said.

"I feel great!"

"You're a liar."

"You're right. What're you doing here?"

"We came to visit you," said Painter.

Pearce said, "You didn't come to visit me, boss."

"*You're* right this time," Baker said. "We're here on a postmortem.

"You found her!"

"In a field by the motorway," said Painter. "Raped. Strangled."

Pearce said, "Please tell them to give me my clothes, Mister Baker! This is no place for someone like me! I can't stand it!"

"No, I told them to take your clothes away," Baker said.

"Will you at least let my folks know so they can look after the dog?"

"Obey your *doctor*," Baker told him. "And rest."

---

It fell on Robin Ashworth that day to behold the cruelest, most ravaging sight this world has to offer: the lusterless desecrated flesh of one's own murdered child.

After he made the official identification and was gone from the infirmary the postmortem began at 6:30 P.M. Chief Supt. David Baker, Supt. Tony Painter and other detectives were present to observe.

Supt. Tony Painter had been the chief inspector of the Police Mobile Reserve during the Lynda Mann inquiry, but had been promoted. A veteran, close to Baker in age and service experience, he was very different in personality. Assertive, often aggressive, he'd tell just about anyone what was on his mind, whether or not he'd been asked. He was tallish, fit, balding, with a smooth unlined face, aviator eyeglasses, and the jaw of a drill sergeant. Tony Painter could fill a room or clear one.

He was the kind of cop who might profanely decry

the profanity found in current films. He'd deal with reporters by telling them nothing they wanted to know while trying to make them believe he had. He'd been weaned in a tough police district in Leicester and risen by virtue of brains and nerve. Not as complex a man as Derek Pearce, he still inspired similar comments: "You either like him or you don't."

The official measurement showed that Dawn Ashworth had sprouted perhaps an inch taller than her parents had realized. Her height was measured in death at five feet five inches.

One couldn't fault the initial assessment that she'd been viciously beaten. The pathology report listed two antemortem abrasions on her upper left forehead, a superficial on her nose with swelling over her left cheek, bruising from the left eye down as far as the jawline, and more from nose to ear, associated with a large conjuctival hemorrhage on the lower left eyelid.

There was swelling of the lips and a linear cut on the inside of her mouth relating to the spring on a dental brace attached to her upper teeth, a brace that was to have been removed on the 13th of August. There were other abrasions on her face and much bruising on the anterior neck at the level of the larynx. On her upper right chest there was a group of abrasions, antemortem.

They found no broken fingernails, and the tongue had been gripped but not bitten. It was concluded that most of the marks on the body had probably been caused by the assailant's attempt to hide it, rather than by the beating that was presupposed, although there had been face, neck and trunk injuries before death, and severe injuries to the perineum where the assailant had viciously penetrated her vagina and anus.

The pathologist's opinion was that she'd died of manual strangulation, and had possibly received a "commando type chop" or suffered some sort of stranglehold where a forearm was pressed against the larynx by an assailant behind her. The pathologist could not completely exclude a ligature. The blows to the side of the face suggested a right-handed assailant, and her mouth lacerations suggested a severe gripping of the mouth to muzzle her.

About the severe injuries to the perineum, the pathologist found that a lack of reaction in one of the abrasions suggested that it had been received at or after death. He found recent hymen tearing and no evidence of old hymen tears. The pubic hair was damp and matted. "The victim," he wrote "was *virgo intacta* and had experienced forceful sexual penetration and acute forceful dilation of the anus as would occur in forceful buggery."

Few of the marks on the battered body of Dawn Ashworth suggested defense injuries. Tapings were taken from various parts of the body, along with oral, vaginal and anal swabs. Fingernail and hair samples were collected and a red fiber was found in the debris covering her. For the record, the cause of death was asphyxia due to strangulation.

The pathologist editorialized in his official report that it was "a brutal sexual attack," a theme that was seized upon by the press, and would, as the rumors and gossip blazed through the villages, suggest lurid stories of Dawn Ashworth's having been raped by bottles and tree limbs and other objects.

The pathology report was not as ruthless and dehumanizing as most. An opinion was mercifully added to the conclusion: "When one considers the amount of bruising in relation to the larynx I have to suggest that the sexual attack occurred after strangulation and, therefore, at or after death." A drop of solace.

An employee who'd worked with Dawn on the last afternoon of her life told police that a boy had bought her a "cuddly toy" and of course that boy was sought, interviewed and cleared. From the joy of being presented with a cuddly toy by an admirer, the vivacious virginal girl had gone immediately to a nightmare death on Ten Pound Lane.

Even veteran policemen sought consolation in the opinion of the pathologist, and hoped she hadn't been aware of all that was being done to her.

———————

Sunday morning Pearce woke up at 6:00 A.M. as he normally did, but he felt anything but normal. Still, he summoned a student nurse and said cheerfully, "I've cracked it!"

"You've cracked it, all right," she answered.

"No, I mean I've beaten it. I'm fine. Magnificent!"

He started hopping down the corridor to show everyone how magnificent he was.

They sent for a doctor and Pearce told him, "I feel a bit of a fraud, doctor. I shouldn't be here."

"You're not going anywhere," the doctor said.

"No, really," Pearce said. "I'm just wasting your time and taking up bed space. I'm signing myself out of here."

"Promise me you won't be completely stupid," the doctor said. "Take *three* weeks off from work."

"Right. I'm going to stay with me mum and dad and rest for three weeks," Pearce promised.

The doctor then said something to the effect that it was Pearce's funeral and left the world's worst patient to his own devices.

Pearce rang the police station and got into another debate with a sergeant who thought it was perhaps unwise to be one's own physician. "I'm giving you an order!" Pearce said. "Get down here with some clothes or I'm going out naked!"

The clothes arrived and Pearce was driven home, weak and sick, his head like a summer squash. He called David Baker at 6:00 P.M. Sunday night, but he was informed he wouldn't be working the Dawn Ashworth murder inquiry. Baker reminded Pearce that he'd worked on previous murders and it was time to let other inspectors gain some homicide experience. Pearce countered by saying he'd at least like to come back and cover the division, in that the doctor had overreacted and now realized Pearce was in great shape.

Pearce showed up at work on Monday feigning perfect health, but regretfully watched the formation of a murder squad to hunt the killer of Dawn Ashworth. He was profoundly disappointed. He *knew* they were also hunting the killer of Lynda Mann.

Headlines were huge. An early *Mercury* edition announced that the missing schoolgirl had been found dead. It was quickly followed by a later edition with a fuller story.

### DAWN: HUNT FOR DOUBLE KILLER

The sex killer who brutally murdered 15-year-old Dawn Ashworth on Friday almost certainly attacked and strangled Lynda Mann, another schoolgirl, whose body was found a few hundred yards away near Carlton Hayes Hospital, less than three years ago.

Detectives hunting the killer, many of whom investigated Lynda's killing, today asked for maximum public help to catch 'a very sick person' and

**said that a tiny scratch on a man's face could be a vital clue to the killer . . . for Dawn put up a brave fight and probably injured her killer.**

There were a few things wrong with that information. First, the police knew that Dawn had been killed on Thursday afternoon or early evening, and second, she had probably not put up a very vigorous fight. But, as in the case of Lynda Mann, it was impossible for journalists, family and even many of the police to believe that the victim of what the press would always call "a horrific sexual assault" would not have battled ferociously.

David Baker's statement to reporters was more accurate: "There is every *possibility* that Dawn put up a struggle and she *may* have injured her assailant, either by scratching or biting him, or he may have injured himself in the struggle."

By the second day of the inquiry the police had a witness from a factory yard across the motorway who claimed to have heard a scream just after 5:00 P.M., quickly followed by another, "like a young child but lower-pitched and muffled." Originally the witness had thought it must be children at play, as indeed it may have been, since he had to have heard it some two hundred yards away, across six lanes of rush-hour traffic. Still, it was the best initial lead to come in.

Because they apparently had a series killer on their hands, the call-out was even larger than it had been for Lynda Mann. More than two hundred officers were assembled.

The Eastwoods, who did not know the Ashworths, were immediately sought out by reporters. Eddie Eastwood obliged them by saying, "We were hoping Dawn would not turn up the same way as Lynda, but

when we heard she had, it was like putting the clock
back. Emotionally, we were back at square one."

And once again, the Eastwoods were scrupulously
honest in describing what they wanted from the law.
"By justice," Eddie said, "I mean imprisonment for the
rest of his life. If we had capital punishment I would
want him to *hang* for taking the lives of these two
children."

Kath Eastwood said, "If someone has been covering
for him, then they're responsible for this second murder and should be brought to trial for it. That person
should consider it and stop him before he does it again.
For God's sake, give him up!"

Just as in the Lynda Mann case, reports began pour-
ing in, hundreds within the first few days. Suddenly
other assault victims were reporting unrelated crimes,
including a young boy who said he'd been indecently
accosted near Ten Pound Lane.

Reporters from all the media prowled hungrily for
impressions of how it was to live in "the village of fear."
A man who usually walked his dogs on Ten Pound Lane
reported that he would always be haunted by the fact
that he had not taken out his dogs that day.

On the third day of the inquiry the police became
drawn toward the "running man," who would prove as
elusive as the "spiky-haired youth" in the Lynda Mann
inquiry. At about 5:30 P.M. on the afternoon of the
murder a woman had had to brake sharply for a young
man who dashed across Leicester Road near the M1
motorway bridge. She described him as blondish, in his
early twenties, of medium height. The time of her
sighting correlated with the report of screams at 5:00
P.M. Another witness reported that a young man had

made a death-defying run across the M1 at rush hour.
The running man became, and would remain, the hot-
test lead.

On August 6th, one week after the Dawn Ashworth
murder, a twenty-three-year-old policewoman, about
the same height as Dawn and with the same slender
build, dressed in a costume replicated with the help of
Barbara Ashworth. She traced Dawn's route from Narborough across King Edward Avenue and up Ten Pound
Lane, hoping to jog the memory of any potential witness. The reenactment of Dawn's last walk was videotaped.

By then, two independent witnesses, one of whom
was a local farmer, reported having seen a man crouch-
ing in the hedgerows on an embankment by King Edward Avenue on the fatal Thursday. David Baker told
journalists, "There is every indication that the man
seen in the long grass near the bridge on King Edward
Avenue and between the crash barrier and the motorway *is* the same person. We also strongly suspect this
man is responsible for Dawn's murder."

And on the day that police asked the public to help
find the running man, Robin and Barbara Ashworth
decided to make their first official appearance at a press
conference. They were told that it could be enormously
helpful to the murder squad and to the public at large.
Moreover, it would satisfy the press. They'd talked it
over and decided to be fair to the police, to be fair to
reporters, to be fair to the public. Robin and Barbara
had spent a lifetime trying to be *fair*, and did so even
now when life had been monstrously unfair to them.

Two police officers, male and female, had been detailed to answer their door, open their mail, and answer
their telephone, in order to shield them from press and
public. The Ashworths were beginning to learn that
murder annihilates privacy. And that the murder of

one's child—destruction of all certitude and continuity—
is the worst thing that can happen to a human being.

David Baker was at that first press conference, along
with other officers from the murder squad, to provide
support and protection. He'd found that, at first, Robin
had more trouble holding together while Barbara appeared resolute, asking questions of the police, wanting
to know *everything*.

For the occasion Robin wore a tan coat, a striped tie,
a fresh shirt, but his dark hair tumbled down across his
forehead, giving him a rumpled look. His face was gray
and he looked several years older to those who knew
him. Barbara seemed composed and cool in a pale blue
frock with a white collar, and a white jacket draped
over her shoulders. But she was pale and dry-mouthed,
her upper lip gumming to her teeth as she grappled
with the smoldering emotions of the survivor. She
seemed to be propping up Robin as they sat before the
battery of reporters. She took his hand in both of hers
and held on to him, as though he might fall.

There were some perfunctory police remarks about
the duty of family members to come forward if they
harbored suspicion about a loved one.

When it was Robin's turn, he said, "No matter what
they feel, no matter what relationship they have to him,
they've *got* to put that aside and do anything to keep it
from happening again."

Barbara's defense at this stage of grief allowed her to
draw on the infinite rage of parents of murdered children. She alluded to the shelterer of the killer as being
as guilty as the killer himself, but then she caught
herself because they were trying to *persuade* anyone
close to the murderer to come forward.

She said, "Lynda Mann's parents have been very
supportive. Obviously, I feel that I don't want to sup-

port any other mother going through the same thing. We've got to find the . . ." She paused and momentarily grappled with her fury, but said, ". . . *fiend*, really, that did this to my daughter . . ." Then she paused again and looked at Robin and squeezed his hand tighter and said, ". . .*our* daughter . . . to stop it from happening again."

Robin broke down once, but caught hold. He said, "I warned her and warned her about the footpaths, but she always assured me she went across the footbridge to Narborough. But children, particularly children Dawn's age, think they know best and . . .and if there's a shortcut on a bright summer's day, for sure, they'll use it."

Barbara Ashworth suddenly looked pale and wan, just before her defenses caved in.

She said, "You know the pitfalls with a child and you obviously try to shield them, but . . . I thought I'd be the *last* person that anything like this would happen to!"

She sobbed then, but Robin picked up for her and resumed the thread of conversation. They both continued bravely until the mob of reporters was satiated.

---

The vicar of Enderby issued a public appeal for the killer to give himself up.

Canon Alan Green, speaking to the Leicester Mercury shortly after hearing that Dawn's body had been found, had this message for the murderer:
You have committed a dreadful crime and you should give yourself up now and beg the forgiveness of Dawn's parents and the whole community.
You should come forward and ease your conscience because at some time in the future you will have to face your creator and account for the terrible thing you have done.

The vicar reiterated that the killer was obviously a sick man, and the vicar, like just about everyone else in the villages, including many members of the inquiry team, continually referred to "good" and "evil," and made appeals to a "conscience." The word "sociopath" was never heard to escape the lips of anyone associated with the inquiry. It was apparently impossible for most to imagine a category of human beings to whom moral judgments of good and evil do not apply. To whom "conscience" or "superego" is irrelevant, because they are simply *without* one.

A professor of psychiatry from Leicester University, when interviewed by a television journalist, touched on the theme of sociopathy. He said, "I think it unlikely that the killer is someone ill in the conventional sense, and *very* unlikely to be someone at the hospital. He may be someone from nearby who no one suspects. He may be regarded by his family as a quiet, even timid man. It's extremely unlikely that his family and friends will believe he could be responsible for these attacks.

"It's likely that he's vulnerable in ways not apparent. His abnormality is in his mind and bursts out only occasionally. Once an episode of violence occurs it becomes the focus of an inner preoccupation and fantasy, and this increases the likelihood of it happening again."

The reporter then asked, "Do you really think this increases the likelihood that it *will* happen again?"

"I'm certain that's so," the psychiatrist answered. "There's no doubt that this type of crime tends to be repeated."

David Baker publicly admitted that extensive inquiries were being made at Carlton Hayes Hospital, but he reassured hospital administrators by saying that the hospital had no patients with a record of extreme sexual violence, and the postmortem revealed a "horrific sexual assault" on Dawn Ashworth.

But despite what Baker said, and despite the opinions of the Leicester professor of psychiatry, members of the inquiry team were being directed to that hospital, whose twin campaniles could be seen from Ten Pound Lane, whose mere presence cast a sinister shadow taller than its giant brick chimney that towered over the villages.

---

Even though the journalists loved to write about police conducting searches "inch by inch"—and to show film clips of "fingertip searches," with gloved policemen on all fours, crawling shoulder to shoulder through the fields sifting debris—the fact is that during any police search, most just go through the motions. It's the same in a Leicestershire village as it is in the Los Angeles inner city, and everywhere in between. The officers, particularly the uniformed officers who get stuck with most of the searching, have lots of things they'd much rather do. They don't believe they're going to find anything, and often don't believe there's anything *to* find. In most cases they're right.

But in any police search, there are a few who actually look and pay attention. Some even make notes and write down names. One of them noted the name of a young kitchen porter who worked at the psychiatric hospital, and was seen loitering around the area of Ten Pound Lane when it was sealed off from the public by streamers of orange tape.

The kitchen porter was sighted more than once, seated on his motorbike, watching with great interest.

# 14
# Confession

At the foot of Ten Pound Lane, by the farm gate on King Edward Avenue, the police parked a mobile incident room to take information from villagers and passersby. A large blue notice board said: MURDER HERE. DID YOU SEE ANYTHING?

After the videotaped reconstruction of Dawn Ashworth's last walk, the murder squad took about two hundred telephone calls, and dozens of people visited the mobile unit.

The most promising new lead concerned a motorcycle that had been parked under the motorway bridge. And there were several reports of a young man in a red crash helmet observed in the vicinity of the bridge, sometime between 4:30 and 5:30 P.M. on the day of the murder.

Most of the dense undergrowth by Ten Pound Lane had been hacked to pieces, leaving splintered tumps and gaping holes in the green tunnel. Still, the missing silver earring was never found, causing the police to wonder if it had been trampled in the field. Or had *he* taken it away, as a memento?

An editorial published as a service to the murder squad, headed KILLER IN OUR MIDST, prompted a spurt of calls over a two-day period:

> It is now pretty certain that we have free in our community somebody who is very, very ill or extremely evil . . . sufficiently ill or sufficiently evil to

sexually abuse and strangle two teenage girls—
girls just like your daughter or the one next door.

Nearly three years ago, an immense amount of
police time and effort, backed by publicity from all
the media, failed to trace Lynda Mann's killer.
Now, it seems pretty certain he has struck again.

It is highly likely that he is local, to Leicestershire if not to Enderby or Narborough. Why
then has he not been caught? Either he has not
been interviewed by police or he has been given
an alibi. In other words, he may be sheltered by a
loving, but misguided wife, girlfriend, mother or
friend.

That person now has another girl's life on his or
her conscience. It is time they made sure that the
killer was put somewhere that he can be treated
or kept away from teenage girls.

The odds are that after Friday's murder he is
marked. If one of the men in your life has a
scratch, a bruise or a cut he has received since
Friday it is your duty to tell the police at once.

If you suspect a neighbour, a friend, somebody
who drinks in the same pub or works with you or
near you then tell the police, in confidence, of
your suspicion.

Catching this pervert is a job for all of us, not
merely the police. For if we don't catch him it
could be your daughter next.

The superintendent who commanded Wigston subdi-
vision, responsible for policing the villages, had a meeting with the Narborough Parish Council and tripled his
normal contingent of two beat officers. The county council agreed to cut back the undergrowth further and
widen what used to be the most scenic shortcut between the villages of Narborough and Enderby. Some
thought it a terrible shame, in that there were rem-
nants of an old Roman road directly beneath portions of
that lovely footpath. But members of the parish council

remarked that in a world growing ever more mad and violent, how could an English village hope to remain exempt?

The next headline produced a rash of phone calls: £15,000 REWARD. NEW BID TO CATCH THE KILLER.

The reward for information leading to the killer's arrest and conviction was offered by a local business-man who asked to remain anonymous.

Hours after it was publicly announced, an event took place that would produce an infinitely more startling headline the very next day.

---

Supt. Tony Painter's murder squad had been assembling some information that actually began to connect with other bits and pieces. Four different witnesses had reported a motorbike. The first saw a red motorcycle parked unattended under the M1 bridge at noon on the day of the murder. A second saw a motorbike parked there at about 4:45 the same day. Another saw a red crash helmet hanging from a motorbike near Ten Pound Lane, that sighting at 5:15 P.M. And yet another witness remembered a motorcyclist wearing a red crash helmet riding up and down Mill Lane on the evening Dawn's body was found, and the next day as well—riding up and back, very slowly, past the Ashworth house.

On August 1st, the day after Dawn was reported missing, but a day *before* her body was found, a police-woman and a detective saw a youth on a red motorcycle in a red crash helmet taking an interest in the search. He was sighted again, in the same spot, three hours later.

And most tellingly, a police constable on security duty at the checkpoint on Mill Lane in Enderby—at

9:20 P.M. on Sunday evening, the day *after* the body was found—was approached by a seventeen-year-old kitchen porter from Carlton Hayes Hospital. The boy was pushing his motorcycle. The officer questioned the boy routinely after the lad volunteered a bit of information.

"I saw Dawn walking up here Thursday night. Toward the gate," the kitchen porter told the policeman.

"Thanks," the officer said. "You'll be contacted in the near future by a member of the enquiry team."

A detective followed it up and spoke to the kitchen porter two days later, when the lad also reported seeing a suspicious boy on a bicycle.

The most astonishing information that crackled through the incident room on Thursday, August 7th, came from another employee of Carlton Hayes Hospital, a friend of the kitchen porter's. This friend had been on holiday the day Dawn Ashworth went missing, he said, but had gone to the hospital to collect his wages. The kitchen porter visited him at 10:00 P.M. the next night and excitedly told him that Dawn Ashworth's body had been found "in a hedge near a gate by the M1 bridge."

When the friend's father overheard the conversation he asked the kitchen porter where he'd gotten his information, for it hadn't been on the telly.

"Someone told me," the boy said mysteriously. "Her body was hanging from a tree!"

Well, she *wasn't* found hanging from a tree, but she was certainly concealed beneath tree limbs and other debris. And she *was* found inside an access gate leading from Ten Pound Lane to the fields, and it *was* just a ten-minute walk from the M1 bridge. And the kitchen porter had this information twelve hours *before* the denim jacket had been spotted!

Still another witness came forth who reported that the kitchen porter, cruising about on his motorbike, had stopped and told him, "Yeah, she was found dead." This, at 1:45 P.M. on Saturday, a few hours after she was found, but nevertheless *before* the press had even been informed.

On Friday, August 8th, at five o'clock in the morning, members of the murder squad drove to the young kitchen porter's home near the Foxhunter Roundabout in Narborough to arrest him.

The boy's father was a gregarious, self-employed taxi driver in Narborough, and his mother, jolly and warm with a welcoming smile for everyone, worked at the Enderby Leisure Center near the home of Dawn Ashworth.

When the police knocked, the mother woke up shouting, "Who's that bumping the door?"

She thought she'd been dreaming at first, but the knocking continued. She got up, put on some clothes and went downstairs. Four members of CID entered the house, informing her that they had to see her older son.

"Is it important?" she asked.

"I'm afraid it is," one of the detectives answered.

"You've not found another one, have you?" she asked.

"I hope not," he answered.

Another detective said, "We've come to arrest your son for the murder of Dawn Ashworth."

"You're joking!" she cried.

"Do you think we'd be joking this time of morning, dear?" the detective asked.

The mother later remembered having to catch the side of the settee in the living room to keep from falling down.

"Where is he?" a detective asked.

"He's in bed!" she said. Then she gathered herself and shakily climbed the stairs, shouting to her husband. "You've got to come! The police want you!"

"Don't be bloody daft!" he muttered.

The first thing he remembered clearly about that morning was getting out of bed and looking at a detective who said, "We're arresting your son for murder."

"You're bloody joking!" he said. "You're crackers!"

When the police woke the boy and told him to get dressed, he said, "Is it about some more questions?"

"Something like that," the detective answered.

Putting on a track suit and tennis shoes, the boy said, "I've got to be at work."

"Don't worry about that," the detective said. "I'm arresting you on suspicion of being concerned in the death of Dawn Ashworth. I must tell you, you do not have to say anything unless you wish to do so, but anything you may say may be given in evidence. Do you understand that?"

The kitchen porter had his father's thick dark hair, but not his chiseled good looks, and certainly not his assertiveness. The taxi driver was considered a controlling parent, but anyone who knew his son might understand the need.

The father watched silently while they searched his son's bedroom. When the tallest detective was on his knees searching under the carpet, the taxi driver cried out, "I *know* my chappie! I know he's not done it because I know my lad!"

The detective answered, "We know him too."

And indeed, the murder squad *did* know a lot already and was about to learn many things about the boy that his parents did not know, and would scarcely believe.

*     *     *

The kitchen porter was driven to Wigston Police Station where DSgt. Dawe and DC Cooke taped the first of many interviews at 8:09 in the morning. The boy sat at a table facing the two detectives and talked quietly. In that first interview he said he'd known Dawn about three weeks and seen her walking about the village. When they asked about his whereabouts on Thursday, July 31st, he said he'd slept in until ten or eleven because it was his day off. Then in the afternoon he'd taken the motorbike for a trial run at half past four, down along King Edward Avenue toward Narborough village. "I were going toward the motorway bridge," he said. "You know where that is?"

"Yes," Dawe answered.

"I looked on the left and saw Dawn approaching the gateway."

"How did you know it was Dawn?"

"By her hairstyle and the way she walked. So I knew it were Dawn," the kitchen porter said.

"Do you know her very well?"

"Just by looks. That's all."

"What was she wearing?"

"A sort of white skirt and a yellow or white jacket. I thought I'd stop and talk to her and ask her where she were going, and that. Then I thought, I've got to get home and do this oil because it might be running out quick. It got to leaking drip drip drip fairly fast, so I just drove straight home."

Dawe asked, "What did you think about her?"

"She were talkative."

"What did you used to talk about?"

"Things."

The sergeant asked him, "Have you ever been with a girl who wants sex?"

"No."

"Never? Never interested you?"

"No. Me dad's warned me," the kitchen porter said.

Then he began rambling on about motorcycles. He said he had another one he was repairing because the bearings were gone. And suddenly he put himself in *another* place on July 31st!

The sergeant interrupted, saying, "You're not telling one hundred percent the truth, are you?"

"I am!"

"Well I'm telling you that you're not."

"Can't remember!" the boy said.

"Well, you said you went straight home on your bike because it was leaking and you'd been doing seventy-five miles an hour, and now you're saying you went somewhere else."

"I honestly . . . I can't remember!"

"Well, you've *got* to remember. It's important to you. I don't think you want to remember. You can't put these things at the back of your mind, or put them out of your mind. Something's happened and you know something's happened. I know you're not telling the truth. Don't sit there worrying about it. What's happened's happened, okay?"

"You're saying that I got the blame for it and that's it?"

"Nobody's blaming you."

Cooke said, "No, but we know that you're not telling the truth, and we *must* have the truth."

"A youth identical to you was seen coming down carrying a red crash helmet."

"It weren't me," the kitchen porter said.

"A bike was seen parked under that bridge," the sergeant said. "If you stopped and had a chat with her, for goodness' sake, *tell* us. Because if you tell lies, even though you may not have had anything to do with it, it makes you look worse. I think you stopped and parked under the bridge. You may well have spoken to her. I want you to tell the truth. It's important to you and it's important to everyone else."

Cooke said, "That's the truth, isn't it? You *did* stop, didn't you?"

The boy replied, "Yeah, I can remember now."

Cooke said, "You *stopped* under that bridge, didn't you?"

"Got under me bike and had a look to see if the oil were coming out."

"And what did you say to her?"

"I just seen her approaching the gate."

"You were holding your crash helmet in your hand, weren't you?" Cooke asked, but the boy shook his head.

The sergeant said, "Friday night at ten o'clock two people say you went round to their house and spoke about the body being found."

"That's what he told *me*! I'll tell you everything he said to me!" the kitchen porter cried.

"You've got to tell the truth," Dawe said. "Something's happened, and if an accident's happened, for goodness' sake, say so. If something went wrong, if you *did* speak to her and something went wrong up that path, it may be that you're not to blame for that. If that's what happened, son, for goodness' sake, tell us. It's important."

Cooke said, "You did, didn't you? You went up that path."

"I went *halfway*, yeah."

"If you tried to kiss or cuddle her, and there was a fight or an argument, for goodness' sake, tell us," Dawe said.

"We've seen it all before," Cooke told him. "You're not talking to your parents now, you're talking to *us*."

Suddenly Dawe said, "The thing is, did you *intend* to kill her?"

"No," the boy answered.

"So what happened?" Dawe asked.

"Right," the boy said, beginning again. "I seen her walking up the lane. Pulled down the side, got off me bike. Started talking to her. I asked her where Queenie was and Michael. She said, 'I don't know.' I said, 'Where you going?' She says, 'I'm going home.' I walked her halfway up the lane and . . . so I walked her halfway up the lane. I goes, 'Will you be all right?' She says, 'Yeah.' So I turns around and come back. Got straight on me bike and goes straight home."

Dawe said, "And what you're saying now . . . what you said in your *first* statement was *not* true?"

"I were worried stiff."

"Why?"

"Thinking I was going to get the blame for all this! I was just upset! Blamed for it! I can't remember what went on that day!"

Cooke said, "I'll tell you why you can't remember. Because you're blocking it out of your mind, son."

"I just can't remember!"

"But are you telling us now that you walked up the path a little way?"

"Only a *little* way."

"On the day that she went missing?"

"No, before! Before the day!"

"Why did you go round to what's-his-name's house at ten o'clock at night and tell him the body had been found hanging from a tree?"

"He told *me* that!"

"All right, but did you go round at ten o'clock at night?"

"Yeah, I went round at ten!"

"Why did you lie a few minutes ago?"

"Because I don't know!"

They stopped the interview then. The kitchen porter was getting so upset he was answering in non sequiturs. He was offered tea and sandwiches.

Supt. Tony Painter arrived at Wigston Police Station along with DInsp. Clancy, and Clancy began with the seventeen-year-old at 11:52 A.M., after once again cautioning him as to his rights.

It was then that the boy began responding to questions about his sexual needs, and told Clancy about a girl called Green Demon, with whom he'd lost his virginity.

"Is it right then that you had pretty regular sex with her?" Clancy asked.

"Yeah."

"Tell us about that, please. Where did that happen? How often? Things like that."

"She said to me that she wanted me to have sex with her. I weren't really interested at the time. She says, 'Here's a chance.' I said, 'All right, I'll take it. Fair enough.'"

"You were only a young fellow then. How old were you?"

"Fourteen."

He had it on a railway bank about three times, he told them. "She were laying on her back and I was laying on top of her. The other times after that she was on her hands and knees. All fours."

"Then what did you do?" Clancy asked.

"I got her twice like that," he said. "I *may* have done

something by mistake. I could have slipped and gone up her bum but I don't know."

"So what do you call doing sex like that?" Clancy asked.

"The phrase we used at school was 'a back shot,' " the boy answered.

He admitted to losing his temper on one occasion while having sex. He tried to explain the temper loss by relating a fantastic tale: "When I were four years old I got thrown into stinging nettles by other kids! I had to go to the hospital and were put in a dark room for two months. . . ."

They weren't sure at this point whether or not the kitchen porter, by virtue of his bizarre answers, was consciously constructing an insanity defense. But on the other hand, none of them had ever assumed that their killer would be a normal everyday bloke.

Then the boy began to respond to questions about Green Demon. He admitted that he'd hit her on one occasion, and that his younger brother had told him he didn't know his own strength. He said that everyone talked about him behind his back and nobody liked him, especially girls.

The kitchen porter said, "I call them slags a lot. Anything that comes into my head: slags, dogs, whores, bitches."

He told them about two girls at Carlton Hayes Hospital whom he referred to in these terms. He admitted to liking girls younger than himself. He talked about premature ejaculation, and how Green Demon would laugh at him when it happened, and how that would make him shout at her.

When they asked him about buggery, he said, "Oh no! I talked to me dad about that and he told me I mustn't ever do it. I watched films about it though, on a friend's video."

But when they put it to him that he *might* have committed buggery with Green Demon, he admitted that *once* it might have happened.

"*May* have slipped," the boy said. "*May* have done it that way. That one time it felt like I put it in a coil of sandpaper. It hurt me and made me bleed!"

With that interview, the detectives had forged a circumstantial link between the killer who'd ejaculated prematurely when he'd killed Lynda Mann, and the one who'd buggered Dawn Ashworth.

The kitchen porter's parents didn't talk to their son from that Friday morning until Saturday evening. They were told that no one could see him prior to that, not until he'd been charged. That until he was charged, he was only helping with the inquiries.

———

Robin and Barbara Ashworth, who were both on compassionate leave from work, were advised by friends and family and even the police to get out and about. For their first outing they decided to go to a market. They needed to get away: A total reconstruction of the murder was being done for a *Crimewatch UK* show for national television, and technicians were at their house.

But when they were alone in the market Robin said it was so strange, shopping without her. Looking at the things *she'd* looked at.

The thought Barbara couldn't shake was, With all these people milling about in the market, why did it have to be *her?*

It was especially difficult looking at earrings. Dawn had had pierced ears and loved earrings.

"I suddenly realized how it must have been uppermost in Dawn's mind to look at those earrings," Barbara said.

Other strange things happened that day. They went to Deer Park and were drawn toward two Airedales at play. They began talking to the lady who owned them and the lady asked where they lived.

"Enderby," Robin said.

"Oh, Enderby!" she said. "Wasn't it terrible about the girl who was killed? Did you know her?"

On that very first contact with the outside world, they didn't know what to say. What *could* they say? The lady looked from one to the other expectantly, and Barbara finally said, "She was our daughter."

"Then we realized that sort of thing would happen for a long, long time," Robin Ashworth later said.

"You don't know whether to let them go on or break it gently to them," Barbara recalled. "That particular lady, we got her name and address. We told her it was all right, that we needed to talk about it. And she did send us a lovely letter."

"People always say, 'We'll be sure to drop in when we come over that way,' " Robin explained, looking toward Barbara. "But we never see them again, do we?"

# 15
# Resurrection

After giving the kitchen porter a rest break, enthusiastic detectives began following up on things they'd been told that morning—particularly about the girl with the CB handle of Green Demon. Supt. Tony Painter injected himself into the interviews at 2:06 P.M. But by then, the boy had back-pedaled and was denying he'd had prior sexual experience with Green Demon or anybody else.

"Me dad would've slippered me if I had," he explained to the man in charge of the murder squad. "And me mum is *very* strict."

He assured Tony Painter that he'd never even masturbated, and when the interview turned to what he'd told the policeman who'd first spotted him during the Dawn Ashworth search, he simply denied most of it. He admitted to having told a constable he'd seen a youth on a bike in the vicinity, but nothing more than that.

When Tony Painter pointed out that the kitchen porter had told the constable a story of having seen Dawn on July 31st, the boy said, "He's just trying to get me in trouble! Then everyone would think *I'm* the one. And they'd pick on me more than anyone. They'd stop looking for him what done it."

When Painter reminded him about visiting a friend and reporting that Dawn had been found hanging from a tree, the boy said, "It was him what told *me* she was

666

hanging by a leg! And he said he heard it from Dawn's older brother. He just wouldn't say it to you chaps cause he don't want to get in trouble."

But Dawn didn't even have an older brother, and his friend's father had verified what the kitchen porter had told him. When these inconsistencies were pointed out, the boy would, often as not, simply respond inappropriately, as though to another question.

He referred to Ten Pound Lane as Green Lane, as did many villagers who lived on the Narborough side of the footpath.

The police had a written statement from a village girl who said, "He followed us around in a strange manner." When the conversation veered to such reports, the boy said, "I follow them around if I fancy them, but I *didn't* follow Dawn up Green Lane."

Sgt. Dawe then asked, "Why didn't you report sooner that you'd seen Dawn?"

He answered, "Me mum said, 'Keep it to yourself. Don't get yourself dragged into it.' That's why."

He was allowed to rest again and at 4:06 P.M. the interview was resumed with Supt. Painter and Sgt. Dawe. And this time things took a dramatic turn.

"All right then," the kitchen porter said to Painter, "I saw Dawn walk toward Green Lane. I saw her walk toward the top of the lane. And I saw a man carrying a stick."

He then gave a rather confused account of walking with her up the lane, and told how he'd noticed the man with the stick following them, later surmising that the man must have waited until the boy was gone before he attacked her.

Asked if, during the walk, he spoke to Dawn about anything sexual, he cried, "Dawn would not talk about sex or even use rude language because she was not like that!"

And then he told of hearing about the search for Dawn Ashworth, and of going home and telling his mother, "Oh, Christ, she's gone missing!" And he admitted saying to himself then, "*I'll* be blamed!"

After that came the most spectacular moment in a long day: Tony Painter showed the kitchen porter a photo of Dawn Ashworth and said, "I think you *were* responsible."

The sergeant added, "I don't think you *intended* to kill her."

And the boy looked from one to the other and answered, "I can't remember. I probably really went mad, and I don't know it!"

"Did you fancy her?" the sergeant asked.

"Yeah, a little bit, but I can't remember any more."

"Describe to me exactly what happened," Painter said.

To which the kitchen porter replied, "She'd gone down and I started putting me hands on her top! She weren't struggling at the time until I put me hands up her skirt. See, I walked up the lane and commenced to touch her bum. And she moved toward me and tripped over a bit. I continued to feel her and she struggled but I held her down. And then . . . then me head started spinning as if I was drunk! I couldn't remember no more until I were running away. There really was no man with a stick. I just went mad! I couldn't help it! Dawn said she wouldn't tell nobody about it! It was like someone else took over! I just went mad! Like it was someone else in me that told me to do it. I didn't want to do it. Someone were forcing me to do it. Making me arms and legs go all over. At first she let me but then she went down. She struggled and me mind went blank.

I don't think it were me that did it. But when I finished and she were getting up I ran off."

But once again, the kitchen porter quelled the detectives' excitement by backing up and denying what they'd just heard him say. "I never had no sex before," he said. "I don't feel in meself that I was responsible for her death. I feel like it's not me that done it. I walked up there and done a few things and said I were going, and then I went. She was about to get up, and that's the last I seen. Apart from being down toward the bridge I were in kind of a trance making me do it. I can't remember I *done* it."

Then he began a rambling, disjointed account of going to his friend's house in Narborough twice on the afternoon of July 31st, and he repeated that his friend had told *him* about a body hanging by one leg.

The interviewers informed the kitchen porter that they'd already followed up with his friend who hadn't even seen him that day.

"He's lying!" the boy said, but then admitted he couldn't actually remember what day it was. He said, "I can only see her walking in Green Lane!"

Then he became angry and cried out, "I never *touched* her! Why should I get the blame? I never even talked to Dawn in Green Lane!"

And suddenly, in the midst of a confession that was confused, disjointed, bizarre, the boy said something eminently sensible: "I want a blood test!"

When the kitchen porter was no longer angry and things were under control, Supt. Tony Painter asked, "If you're now denying everything, why did you say the things earlier?"

"To settle your story," the boy said sullenly.

Then they replayed the earlier parts of his confession where he'd admitted attacking Dawn Ashworth.

"Did you say these things?" Painter asked.

"Yeah, I should've done what me dad said. Keep it quiet because I've not done a thing."

"Well, did you or didn't you tell a uniformed officer on Sunday night that you'd seen Dawn walking to the gateway of Green Lane?"

"I talked about . . . I discussed it with me father," he said.

And at the mention of his father the boy suddenly threw himself across the table and started to cry.

———

The police, who were now trying to check out every utterance from the mouth of the kitchen porter, discovered another young witness who'd watched the police searching a field on August 1st, the day before Dawn's body was found. The young man said that the kitchen porter had ridden up on his motorbike and casually remarked that the police should search the culverts by the M1 motorway. Another witness had seen him three times during that day of searching. He seemed to have been everywhere that afternoon. Watching.

During one of his more incoherent statements to police, he mentioned being with another friend on Thursday during the hour when Dawn was murdered, and the next day as well. But the friend was contacted and told detectives that he had not seen the kitchen porter that day or the next.

When the interview was resumed early Friday evening, a question was followed by a suggestion. "You've told us you walked up the path with her. You were laughing and joking and touching her up and she didn't mind it. You don't know exactly *what* you've done, do you?"

Once again answering unresponsively, the boy replied, "Not all night. I was probably in the garage."

Tony Painter said, "You weren't probably in the garage."

They were all getting tired, impatient, frustrated. The sergeant said, "It's not *probably*. It's not *probably* at all."

"I don't know! I wasn't there, was I?"

"You *were*."

"I know meself I *weren't* there!"

"You've told us twice that you *were* there," Painter said. "You *were*. Now come on!"

"Was she alive when you left her?" the sergeant asked.

"I don't know."

"Did you panic?" Painter asked.

"Yeah."

"Why did you panic?"

"I don't know. I just did."

"Why did you panic? Wasn't she moving?"

"No."

"Why wasn't she moving?"

"I don't know."

Then Painter asked, "What did you do to make her *not* move?"

"I think it was when I laid down on top of her," the boy answered.

"Where were your hands?"

"On her arms."

"And what were you doing to her when you were laying on top of her?"

"Just had a laugh and a joke with her. I said to her, 'I ain't going to let you go.' She just started laughing. She was crawling all over me."

"So what did you do?"

"Moved up towards her face and sat on her chest and that's what done it. Sat on her chest."

"Was that before you hurt her or afterwards?"

"That were before."

"How did you feel when she stopped moving?"

"Dunno."

"Come on, tell us *more*."

"When I realized she weren't, I thought, Oh shit, oh Christ! I just got up and went back down the lane. I thought she had a heart attack or some like that."

"Where did all this take place?"

"Near the hedge by the ditch. I can't remember because I know I didn't do it. That's why I can't remember for certain."

Sgt. Dawe said incredulously, "You say you didn't do it? You don't think you *killed* her?"

"I can go up the lanes. . . . I don't even know what happened!"

"But you were just telling us a few seconds ago that you sat on her chest!"

"I was telling what I *would* have done!"

Later that night it continued. There was more disjointed talk about "feeling" her and then the kitchen porter said, "I'll have a try telling you. If I can remember."

"Go on then," he was told.

"Well, I got as far as putting me hands up her top. Then I put me hand up her skirt. She said no. I forced. She started shouting. I put me hand over her mouth to shut her up. Put me hand down her pants. She wouldn't let me. She turned her head over. . . . It's all I can remember. Then she were lying there still. I just pressed really hard on her mouth with me hand over her nose

and her mouth. She suffocated. That's all in my memory. I couldn't leave her where she was so I hid her."

"How did you hide her?"

"With a load of brambles underneath a hedge."

"How did you leave her? In what sort of position?"

"On her front."

"What? Lying on her stomach or her side or her
back?"

"Side," he said.

And that was more or less how she was found.

"Where was this load of brambles then?"

"It was on the Green Lane path."

"Did you have to pick her up to hide her?"

"No."

"Are you sure?"

"Yeah, sure."

Painter said, "Son, you've told us so much. Just
continue."

"I'm trying to tell you the truth! I'm trying to prove
to you that it's not me that done it. I don't even know
where she was hid. I said to that officer, 'Point out
where she was found' when I pointed to that gate to
find out where she was found."

At 9:37 P.M. they put on a new tape and again cautioned him as to his rights.

And Painter said, "You were telling us what happened."

"When I put me hand up her skirt she were shouting
and screaming so I put me hand over her mouth to shut
her up."

"What did you do after that?"

"So I thought, 'I've got to put some mark on her like
she were strangled.' So I did. She were strangled."

"What sort of mark was that?" Painter asked him.

"Grabbed her round the throat and squeezed her
dead hard. Pressed her about there. Really hard."

"What else did you do?"

"It's all I can remember doing."

"So you were sexually excited, were you?"

"Yeah."

"You'd got an erection, yes?" Painter said. "What else did you do?"

"I don't want to say."

"You've told us everything else, son."

"I lifted her skirt right up, took her drawers off and I had sex with her. That's all I done."

"Then what did you do?"

"That's when I moved her. That's when I moved her into that deep undergrowth. I carried her over the fence to the field where I hid her."

"How did you knock her down?"

"I put me feet behind her and pushed her."

"A few minutes ago you said that you'd already gone through the gate into the field."

"It was a mistake, that was."

"Where did you do all this? In the lane *or* in the field?"

"In the gateway," he said, offering a compromise.

"When you say you took her pants off, did you take them right off?"

"Yeah."

"What did you do with them?"

"I don't know. Just chucked them away."

"But you've just described how you put them back!"

"I put everything *else* back. I put her skirt back on. Bra back on. Shirt. Tucked that in and that's it."

"Have you anything else to tell me?" Painter asked.

"No. Nothing at all."

"How were you rough with her?"

"I don't know."

"Just *tell* me."

"I hit her."

"What did you hit her with?"

"That," he said, showing his clenched fist.

"Where?"

"In the face. It was here, round the chin."

"Around the chin?"

"I think so, yeah. I hit her in the mouth."

"Did you do anything else?"

"I just hit her three times."

"Was that before you indecently assaulted her?" Painter asked.

"Yeah."

"Was that because she didn't want to do it?"

"Yeah."

"What else did you do?"

"That's about all I did. Hit her. Kicked her a few times. That's about all."

"Kicked her a few times, did you? Where did you kick her?"

"In the ribs and that."

"What made you do what you did?"

"What? Hit her?"

"In the field. Just tell me what made you do what you did."

"Just cause I liked her at the time. I wanted someone to have sex with and I didn't think she'd let me so I tried it on. All right so far. Then she started panicking so I thought, If I leave her she'll tell her mum and dad and I'll be in trouble. So I did something about it. She started screaming so I put me hand over her mouth and with me other hand I fingered her. I took her pants off and had sex with her and buried her and that's all I remember doing. I walked straight back down the lane."

"What time did you get home?"

"I got home about five."

"Anybody see you?"

"Me mum seen me come in."

"Is there anything else you want to tell us about it?"

"Nothing else."

"How do you feel?"

"Not very good. I feel bad that I done something I shouldn't have done." Then he added, "I'm not quite sure what I done."

"What made you have intercourse with her?"

"Because I had an erection. I wanted to get rid of it somehow so I wanted to find out what it was really like so I done it."

"Normally?"

"I *think* so. There ain't any other way else to do it!"

The police quickly contacted more witnesses who could testify to the peculiar ways of the kitchen porter.

A young woman told them that one night in the Red Lion Pub he had walked up and said, "I wouldn't half like to fuck you," and tried to put his hand up her skirt.

And finally, a woman from Carlton Hayes Hospital reported that the kitchen porter had shocked her by saying he was the last person to see Dawn Ashworth alive. And the woman had noticed that he had some scratches on his hands.

He said to her, "If they found the body could they revive it?"

At first she thought he was joking, but he was such a thick sort of lad that she bothered to assure him that resurrection was not possible.

"What would happen if they *could?*" he asked her, no doubt with his sly little smile.

The headline announced it boldly:

DAWN: MURDER SQUAD POLICE ARREST YOUTH

———————

The grandparents of Dawn Ashworth took the girl's death *very* hard. Barbara Ashworth's mother was never able to talk about it and her father couldn't talk about it enough. Her mother told Barbara that she'd stopped having periods on August 10th, the day after the arrest of the killer was announced.

And Barbara reminded her mother of the time when Dawn, less than five years old, had said to her grandmother, apropos of nothing: "I won't know you when I'm fifteen."

Since Barbara's parents didn't live locally, they indeed had not seen Dawn since her fifteenth birthday on June 23rd, and now never would.

To console her mother, Barbara said, "Perhaps children have a way of foreseeing events. Perhaps it was inevitable."

Parents of murdered children quickly learn that all they have for barter and trade is a bit of solace.

# 16
# Beyond Imagination

When he'd finished breakfast on Saturday morning, after he'd been locked up at Wigston Police Station for twenty-four hours, the kitchen porter was presented with a typed statement for a legal endorsement and signature. Sgt. Mick Mason, hoping at last to be able to tell Kath Eastwood that her daughter's killer had been caught, was there at Wigston when Supt. Painter concluded for the record that the kitchen porter's knowledge of the crime "went far beyond imagination and was consistent with facts."

The seventeen-year-old had decided to clear the air, even to his buggery of Green Demon, as described in the document. The boy read the statement, nodded, and said, "Yeah, what's in that paper there, I did. I *did* go up her arse."

And since the boy customarily dropped his h's, Tony Painter thought he'd said "ouse" and asked him, "Which house was that?"

The kitchen porter pointed to the document and said, "Up her *anus!* What you're saying here!" Then he added, "She didn't *mind* it none."

The boy's parents were allowed to see him Saturday evening. After talking to him, they tried to tell all who would listen that their son was a bit simpleminded and couldn't have killed anyone.

His mother offered an alibi: "*Heidi* were on TV that

179

Thursday. I got talking to my friend and she said her kids always watch it. *Heidi* comes on at quarter to five and stays on till quarter *past* five. My son were sitting in our house watching *Heidi*. Susan, my brother's daughter? You can talk to her. *She'll* tell you!"

She later added, "He came in the day the Dawn Ashworth stuff were on the telly, and he said, 'I seen her. I seen her go cross the road to Green Lane.' 'Are you sure?' I asked him. 'I were testing me bike,' he said. 'I seen her. I'm going to the incident room.' I said, 'Look, you got to be sure. You can get yourself in a lot of trouble.' I said, 'You keep away!' "

"He went to work and somebody at work told him to go down to the incident room," his father explained to the police. "They told him it could help. And he *did* do. I think it were the reward. Somebody fingered him. And anyway, you're looking for a lad with *blond* hair, the one that ran across Leicester Road and across the motorway. Not my laddie!"

It hadn't happened exactly the way the kitchen porter's parents thought it had, and the police weren't about to give them details of what had been said in the confession. And of course the police were no longer looking for *anyone*.

A detective listened to the parents politely and said, "You should call a solicitor as soon as possible."

"We don't have no solicitor," the father replied, and the detective gave them a list of four law firms.

One of them was familiar, having represented the boy's grandmother in a dispute with a neighbor. When the kitchen porter's mother mentioned to the detectives that they wouldn't be able to reach a lawyer on Saturday evening, Supt. Painter said, "Well, I know somebody who works for that firm. I know his personal number. I can ring him for you."

The man to whom Painter referred was Walter Berry, the same solicitor who had represented Eddie Eastwood in his bankruptcy problems.

Painter told the parents, "If there's anything I can do for you, let me know. He'll be in Magistrate's Court on Monday."

At 2:00 A.M. Sunday morning, while lying awake in bed, the kitchen porter's father broke into a sweat. "It hit me like a brick on top of me head!" he later said. "Monday morning there's going to be a lot of problems down at the court! They caught the bloody murderer as far as the public's concerned! Our name and address was in the evening paper already!"

Late Sunday morning he managed to contact Tony Painter by telephone and said, "There's going to be a hassle down at court!"

"Possibly," he was told.

"What help can you give me?"

"I can assign two officers to be your bodyguards," Painter told him.

And true to his word, he sent a pair of detectives.

"They was two of the biggest blokes you could imagine," the father recalled. "Both family men with girls and lads. What a grip they had when they shook hands! Wouldn't need handcuffs, those two."

"Ever such nice chaps," the mother said. " 'You get any problems, phone calls, letters, give us a ring,' they said. 'You want some shopping done, we'll do it.' "

"Of course we always wondered if they were told to write down anything they heard in our house," her husband said. "Still, you couldn't fault those two."

During his last conversation with Supt. Tony Painter, the seventeen-year-old decided to come clean and confess *all* his crimes. It could be that Painter had sensed the boy had something to add, because he said, "Now, son, is there anything *else* you'd like to tell me while we're here? I'll gladly listen."

"I should tell you about something that I was *forced* into doing," the boy said.

"Well, you tell us," said Painter.

"This were about six or seven months ago. . . ."

The kitchen porter then described an event that was quickly investigated and verified. He told about a young girl who he said was eleven years old, but who detectives would learn was only nine. He described how she and he had been together watching a teenage couple kissing and cuddling and how the young child had made advances to him.

"She hopped straight on me bike. She were rubbing me up. Getting jealous. So I started to rub *her* up. Me friend was fingering his girl, but I couldn't finger this one. She's too young. Can't do it. Then she made this big commotion, yelling to the others, 'He's got this thing up in me!' She were *shouting*."

"What did you do?"

"I just felt her up outside her pants. I didn't go down inside! I didn't *want* to touch her! I just had no choice!"

"You got carried away, did you?" Painter said.

"If I refused, she'd hit me!"

"I see."

"She'd kick you!" the boy said.

Painter said, "Tell me this: Has it happened more than once?"

"Yeah, twice."

"Are you sure?"

"Yeah."

The police quickly found and interviewed the child, and got the story of the kitchen porter's putting his fingers in her "money box."

They'd also had Green Demon examined by a physician and got signed statements from her as to the boy's proclivity for buggery when he was fourteen years old.

"We were praying for somebody to come in and speak to us," the kitchen porter's mother later said. "We wanted to tell them our boy didn't do it. We just wanted to tell them *anything*. Later, when people did talk, they said they'd *wanted* to come, but they just didn't know what to say. A few crossed over the street when they had to pass by our house. It was the same as when somebody dies. You just don't know what to say."

"I swore I'd never work again if my laddie went inside," her husband said. "If they did that to us I vowed they could bloody well support us on the dole!"

Like their son, they felt imprisoned. The two policemen detailed to look after them came frequently to see if they were all right, and to provide their tenuous contact with all of "them," those omnipotent minions of British law who, the parents believed, had stolen their son.

———

That Monday, the kitchen porter was in the dock for three and a half minutes. He answered yes a few times, and that was all. He was remanded to police custody for seventy-two hours. His defender, Walter Berry, raised no objections and made no bail application.

The boy wasn't his usual "scruffy and mucked-up" self, as his mother described him. He'd combed his hair and wore a buttoned shirt and proper trousers and a black corduroy jacket. He was flanked by two uniformed policemen at all times.

As to how his family fared, things were both better and worse than village gossip had it. There was no "hassle" at court, and the village rumors that their house had been stoned were unfounded. They *did* receive several phone calls like those received by the Ashworths when Dawn was missing. They'd pick up the phone and be met with silence. They told their police bodyguards about it, and the next day the police arranged for an ex-directory telephone number.

They received only one hate letter and it was anonymous.

Like the Ashworths and Eastwoods, the kitchen porter's parents isolated themselves during that time. For the first three days they didn't eat at all. Then they were put on medication: five tranquilizers a day and three sleeping pills at night. To shop for food for their younger son, they went to a butcher shop in nearby Hinckley. It was there that the kitchen porter's mother suffered her first anxiety attack.

"I couldn't reckon the money out," she later explained. "I couldn't make change! The item was ninety-nine pence and I couldn't count it out! The man had to help me. Then I went into another shop and bought the same thing twice. It were just as though I got word-blind during those first days. I'd look at a thing and couldn't make it out. I'd put water into the kettle without a tea bag."

"We was both on the verge of a nervous breakdown," her husband related. "I couldn't drive me taxi no more, but being a self-employed driver I couldn't claim unemployment. The taxi firm stood by me all the way, though. They even took up a collection at work."

As the weeks wore on, there were rumors that the family had been driven out of their home and had moved to another part of England. And it was generally believed that the father had been fired by the company from which he operated his independent cab. The newspaper printed a report that he'd had to leave his job because of threats.

None of it was true. The fact is they stayed put in Narborough, and visited their son in Winson Green Prison at Birmingham, a high-security facility where he was being held while the police prepared the case.

The kitchen porter's younger brother was about the same age as Dawn's younger brother and knew Andrew Ashworth slightly. One or two boys had reportedly made vague threats against the kitchen porter's brother but that was all.

"Two of me mates said, 'You stick around with us and nobody's gonna tooch ya,' " the brother later said. And nobody did.

When police came to their home in Land-Rovers to take away the kitchen porter's motorbike for forensic work, the family was verging on paranoia. "I think there's something wrong with our phone," the father said, after ringing Tony Painter. "Are you tapping our phone?"

"You have to go to the Home Office and get written approval to tap somebody's phone," Painter told him. "You need a bloody good reason. In fact, a reason involving national security, and this case hardly applies."

The cruelest of all for the kitchen porter's family was the alienation.

It was similar to that reported by the Eastwoods and the Ashworths. The alienation and loneliness of victims.

———

Four weeks after she was taken from a village footpath, Dawn Amanda Ashworth was buried in the little cemetery behind St. John Baptist Church in Enderby, after a simple service for the family. The vicar described her as a "bright, lively, charming young lady,

obedient to her parents, loyal to her family and full of
the joy of life." They sat in the old granite church,
honeycombed with wine-yellow light streaming through
Gothic arches, and smelled candle wax, flowers, old
hymnbooks, mortality. And *tried* to fathom what cannot
be fathomed—chaos, caprice, discontinuity.

Like Lynda Mann, Dawn was buried in the village
where she'd lived, and just a few minutes' walk from
the footpath where she'd died.

Two hundred showed up at the cemetery behind the
stone church to bid farewell to a girl who'd seemed to
be acquainted with everyone in the village. Dawn's
parents and brother, dressed in subdued grays and
blues and black, followed behind the vicar and server
clad in cassocks. They walked through the churchyard
to the graveside piled high with wreaths, including one
from Lutterworth School that said: "We love you."

Barbara carried a long-stemmed rose and after they
lowered the casket she kissed the flower and dropped it
into the open grave.

As was the case at Lynda Mann's funeral, detectives
came, but this time only to pay respects. Not to scan
the crowd of people who waited for the family mourners to depart before they filed past the open grave. Not
to look for a killer.

A lesser misery the Ashworths had to endure was
trying to cope with condolences to parents of a murdered child—condolences no one knows how to offer,
and no one knows how to receive. After the funeral,
Robin refused to go near that churchyard. Barbara was
able to tend Dawn's grave by constantly reminding
herself that her daughter was not really down there.

Afterward, when her husband would make any sort of
tentative sexual approach, Barbara reported that she'd
"almost crack up." In describing it, she said, "Then I'd

think: That's all *he* wanted from *her*. Why not let her walk away afterwards if that's all he wanted?"

Even when a more normal relationship gradually resumed, she'd be in tears afterward, still thinking, That's *all* he wanted from her! Why not have it and let her walk away? He must've done others and let *them* walk away! Why not Dawn and Lynda?

"It was something I could never get over," she said. "Or if that's *all* he wanted, why not go to a prostitute? Or if he *had* to rape, why not just let her walk *away?* Because I could have seen her over that. There would've been mental scars as well as the physical, but she'd still have been here. I could've seen her through it."

When Chief Supt. David Baker and Supt. Tony Painter invited the Ashworths to police headquarters, Baker explained that they could at least be consoled by the fact that their daughter's killer had been caught, would undoubtedly be convicted, and would be imprisoned for a long time.

Robin Ashworth said, "You're sure then? You're *sure* you've got the right one?"

"There's virtually no doubt," Supt. Tony Painter said. "We're convinced."

But Barbara Ashworth still lived in torment, dwelling on Dawn's last moments on earth. "Your imagination runs riot as you tend to imagine what she went through. The only real consolation I got, if you can call it that, came from knowing a lot of what he did to her he did at the point of death. Which means she didn't know about it."

---

As the case against the young kitchen porter was meticulously assembled, it became apparent that he, in the words of one detective, was "thruppence short of a pound." Another detective who'd investigated many reports about the boy's bizarre behavior called him "the flippin village idjit you always hear about."

Sgt. Mick Mason, who still visited the Eastwoods, said to Kath, "Yes, well, he *must* have murdered Lynda too." But then recalling his own statement—"Lynda would've been able to sort out a fourteen-year-old"—he may not have been entirely convincing to Kath and Eddie Eastwood when he assured them that Lynda's killer had been caught.

There were a great many attempts during the hours of interviews to break down the kitchen porter's stubborn insistence that he did *not* have anything to do with the murder of Lynda Mann. As incredible as it seemed, what with the identical *modus operandi*, it had to at least be considered that there *could* have been two separate killers in the Mann-Ashworth murders.

Of course, the seventeen-year-old had been given a blood test and found not to be a PGM 1 +, A secretor, but the forensic scientists who had tested the semen stains were dealing only in "probably's" and "maybe's" and seldom could exclude anyone "positively." It would have been hard to find a detective anywhere who would stake his reputation on something as iffy as blood typing and grouping.

As far as the police and public were concerned, it was a matter of tidying up. It was reckoned that the speedy inquiry into the murder of Dawn Ashworth had cost £113,000 in overtime payments, a report of which was made to the county council. The incident room was closed and officers were returned to regular duties.

Undergrowth at Ten Pound Lane was to be drastically cleared, and the path opened wide, never again to be the secluded pastoral footpath where leaves brushed your face on a bright summer day.

---

What prompted the next move is open to debate. According to the kitchen porter's father, he asked the head of the inquiry if his son had had a semen test and was told it wouldn't be needed.

"Then I got to thinking," the father recalled. "I'd read somewhere, maybe in *Reader's Digest* or *Tomorrow's World*, about this DNA testing that the chappie in Leicester had discovered. I told my laddie's solicitor to look into it."

The solicitor later reported to the family that Tony Painter hadn't read anything about DNA testing, but promised to check it out.

The police version is that, unprompted by a solicitor or anyone else, Chief Supt. David Baker had decided to try the new technology in order to make a case against the kitchen porter for the murder of Lynda Mann.

However it came about, the semen sample from Lynda Mann and some blood from the kitchen porter were delivered to a young geneticist at nearby Leicester University who claimed to have come up with a wondrous new discovery called genetic fingerprinting.

# 17
# The Window

Derek Pearce wasn't an easy man to pity. He often berated subordinates in the presence of their peers. It wasn't uncommon to hear him barking something like "Don't be a lazy twat! Pull your finger out!" Eighty years ago he'd have carried a sword cane.

But there was another Derek Pearce behind it all— the torchbearer. If one of them ever mentioned the ex-wife, he'd clam up. Once he was heard to say, "It's a yacht-club kind of life she lives in Hong Kong. But she's *all right*." There was more than a note of regret in his voice when he added, "No woman could mean more to me than my job."

When the kitchen porter got arrested, Derek Pearce was "on division," engaged in ordinary police work, but he tried to keep in touch with what the Dawn Ashworth murder squad was doing. He listened to the recorded confessions of the kitchen porter when the lad was, in Pearce's words, "eerily toing and froing" as to whether he had or had not been in Ten Pound Lane when Dawn Ashworth was murdered, and had or had not *committed* the murder.

Pearce envied Insp. Mick Thomas and the others for being able to detect and arrest the killer of Lynda Mann, the failure he'd never gotten over.

---

During most of 1986, Dr. Alec Jeffreys had received great honor within the scientific community and he'd helped to make a bit of legal history by proving, in a highly publicized lawsuit, that a French teenager was the true father of an English divorcee's child. His continuing work in deciding paternity for immigration disputes had brought him a degree of attention that was disrupting his research.

Still, he hadn't had the kind of high-profile forensics case that excites the imagination of the public at large. He was quoted that summer as saying, "It is a perpetual struggle trying to get funds."

Whether or not he was anxious for a famous forensics case, he was about to get one that would put his face and name into news stories throughout much of the world.

Jeffreys was asked by a detective inspector from the Leicestershire Constabulary to analyze samples of blood and semen to assist in the prosecution of the confessed killer of Dawn Ashworth. The police hoped to prove that their killer was also the slayer of Lynda Mann.

Having read a great deal about the horrific murders, Jeffreys eagerly accepted, and in September he analyzed the rather degraded semen sample from the Lynda Mann inquiry. During the final stage of the process, he studied the radioactive membrane with the DNA on it. "And there," Jeffreys later recalled, "we could see the signature of the rapist. And it was *not* the person whose blood sample was given to me."

The next move was obvious: A sample from Dawn Ashworth had to be obtained and tested. Jeffreys was forced to wait a full week before he was able to pick up enough radioactive material from the Ashworth sample. Jeffreys called it "a nail-biting week" because he'd been

virtually assured from the beginning that the same rapist *must* have killed both girls.

When the plastic film came up, he studied it and
rang Chief Supt. David Baker's representative "at some
dreadful hour."

"I have bad news and good news," Jeffreys told him.
"Not only is your man innocent in the Mann case, he
isn't even the man who killed Dawn Ashworth!"

Jeffreys said that the detective's first response to that
was not repeatable.

Finally, the detective said, "Give me the bleedin
*good* news then!"

Jeffreys told him, "You only have to catch one killer.
The same man murdered *both* girls."

As quickly as he could assemble reinforcements, David Baker rushed to Jeffreys's laboratory at Leicester
University. His entourage included a forensics scientist
from the Home Office.

Jeffreys pointed to an X-ray photo and said to those
assembled, "This is the genetic fingerprint of Lynda
Mann, which we found in the stains composed of a
mixture of semen and vaginal fluid. We can compare
this to the DNA from her own blood sample and see
her genetic characters, as expected."

Jeffreys again pointed to the picture, which looked to
the cops like nothing more than the bar codes on a box
of washing powder. "And then we see two more. The
last profile is of the assailant." Jeffreys stopped pointing
and said, "There are two bands so it means there was
one man. If there were two men involved in the attack,
there would be four. Three men, six bands or stripes.
Terribly simple, if you understand it."

Well, sure. To men who'd done "old-fashioned
bobbying" for nearly as long as Jeffreys had been on
earth, with little enough help from people like him with

his bloody Oxford accent and his hand-rolled fags, to these coppers who'd put more hours and sweat into the Mann and Ashford inquiries than had gone into any others in the history of Leicestershire, it was "simple," all right. It was simply bloody outrageous!

Jeffreys went on. He put up another X-ray photo and said, "This is the genetic profile of Dawn Ashworth. There are two semen stains taken from the vaginal swab and from the clothing stain. It shows two bands not attributable to her blood sample, but it's the same as found in the semen stain from Lynda Mann. First conclusion, both girls were raped and murdered by the same man. Second conclusion, your man isn't the killer. We have here the signature of the *real* murderer!"

"We couldn't challenge it," David Baker said later. "How do you challenge brand-new science? Nobody else in the bleedin world knew anything about it!"

The Home Office forensics scientists were to be trained in Jeffreys's method from the technology he'd passed on to them. In a few weeks they'd be able to verify or challenge Jeffreys's conclusions, Chief Supt. David Baker was promised.

"*Any* chance of a mistake?" was about all Baker could manage, at the end of a long and painful debate with the intractable geneticist.

"Not if you've given me the correct samples," Jeffreys answered.

And that was that. The dazed policemen lurched out of Jeffreys's office. Hanging from the notice board outside the laboratory was a copy of a letter that couldn't have amused them at the time.

Dear Sir, Madam:
    We are interested in the new genetic finger-printing recently devised by Leicester Univer-

sity. As owners of self-catering accommodation we wonder whether this method could be used to verify persons responsible for leaving urinated beds.

We enclose an SAE and look forward to your reply.

With thanks,
Yours
faithfully,

The kitchen porter was scheduled to go to court in late November, three years to the day since the murder of Lynda Mann. But something unusual happened the night before, something the boy couldn't figure out. In the high-security cells they always left the red light on all night so they could make periodic checks to assure that no one was attempting escape or suicide. That night they turned it off for the boy. The next day he was taken to Charles Street Police Station in Leicester and treated to sandwiches. He was transported *without* handcuffs.

The parents of the kitchen porter received a telephone call from their son's solicitor who said, "The whole thing's blown up in their faces. I've been told that *unofficially*."

There were only a few people present in the gallery at Crown Court, Leicester, on November 21, 1986, when legal and forensic history was made. The seventeen-year-old became the first accused murderer in the world to be set free as a result of the DNA test known as genetic fingerprinting. The boy's solicitor drove him from the courtroom to his parents' home after a hiatus of three months and ten days.

The Leicestershire Constabulary called a news conference that afternoon. Asst. Chief Constable Brian Pollard, in uniform, sat at a conference table before a room full of reporters. Chief Supt. David Baker sat on his right and Supt. Tony Painter was next to Baker. All three looked like they'd just been told that Libya was moving its London embassy to Leicester.

Pollard, facing batteries of lights and cameras, donned half-glasses and read a *very* carefully worded introduction from a prepared statement regarding the kitchen porter and the DNA tests. Then he quickly concluded by saying, "Those tests did not implicate him in the murder of Lynda Mann. Further tests were asked for by the investigating officers in the murder of Dawn Ashworth. The tests were carried out and results were checked by scientists at the forensics science laboratory in Aldermaston who confirmed the findings. The result of the test indicate that a person *as yet unknown* was responsible in the deaths of both girls. The Crown Prosecution Service have been kept informed and have decided that proceedings should be discontinued."

When questions were allowed, one of the first went straight to the commander of Leicestershire CID, who was asked when it was that they realized they'd "committed a blunder."

David Baker, with a look that could have radioactivated every chunk of DNA in the room, said, "The Leicestershire CID did *not* commit a blunder."

Then he tried to turn down the rems by explaining that the kitchen porter had been charged only after he'd confessed during a tape-recorded interview session, part of which had been conducted in the presence of his lawyer.

Baker cryptically added, "He is not responsible for certain *aspects* of that murder."

"Has he been totally eliminated?" a journalist asked.

"*No one* has been totally eliminated," the chief superintendent answered.

The television news that night began with a startling lead: "Police have reopened the hunt. Youth accused of murder is freed."

Another network newscast began, "Father says he should never have been arrested. Police go back to the drawing board."

Still another said, "Police are no nearer today than they were three years ago."

The next day the headline was huge:

### YOUTH GOES FREE
### NOW PROBE IS ON INTO DOUBLE MURDER

Both Chief Supt. David Baker and Supt. Tony Painter spoke to the kitchen porter's parents after he'd been discharged by the magistrate at Crown Court.

David Baker told them, "We know what you've been through. You think we don't, but we do."

When Tony Painter spoke to the boy's father, the unemployed taxi driver said to Painter, "A few weeks ago I could have willingly killed you."

"I can understand that," said Painter.

Both police officials were exceedingly polite and patient as they relayed confidential information they'd learned from the boy.

The father made a statement to journalists who came to his home on the day of his son's release. He said, "The lad is staying with his grandmother for the next few days but we'll have a family get-together on Tuesday."

When asked about his feelings toward the police he said, "They questioned him for fifteen hours, and although I believe they followed procedure, he was bound to be confused after all that time and just said the things they wanted to hear."

As to other legal implications, he said he didn't want to comment until he'd met with his son's solicitor.

When the father was interviewed by a national television news team, he said, "We have no idea what legal course of action we may take." He would not let anyone interview his son personally, but described the boy's reaction by saying, "The furrows down his cheeks show tears of relief."

Villagers, reporters and the public at large were astonished at how magnanimously the family was behaving. There had been no threat of a lawsuit for wrongful arrest, no demand for compensation.

Three days after the boy's release his solicitor, Walter Berry, told the *Leicester Mercury,* "They are taking no action whatsoever. They think the police acted absolutely superbly throughout. They have no complaints about the police."

It was indeed unusual to hear a criminal lawyer in the role of police cheerleader, so Berry was asked what advice he'd personally given to the family. The solicitor declined to state, saying it would be "improper." However he added that the family was "completely satisfied with the way police had handled the case."

Actually, the kitchen porter's parents had been given a transcribed copy of portions of their son's lengthy confession. After they'd recovered from the shock and dismay of learning about numerous sexual incidents— especially about a prosecutable indecent assault on the nine-year-old girl—and after understanding that their son was *not* being charged with any of these offenses, everyone perhaps decided that a quid pro quo was in order.

\*     \*     \*

The seventeen-year-old did not return to his job at the hospital, but stayed at home every day with his parents. His mother voiced concern about his lying around the house as though still in prison.

"He's going to get a bit funny," she told her husband, perhaps unable to admit that her son had always been a bit funny.

"He were frightened to go out the house," his father said. "So we thought a pet would be a bit o' company."

They bought him a nectar parrot, raven black with a tangerine mask. The boy obviously loved his little carnival bird and called him Dusty.

When their son did go out, one of them would accompany him everywhere he went. They were afraid there were people out there who still believed their son was the footpath murderer, people who would "get him."

"I'll always have me mum or dad with me," he said to a television newsman when his father allowed one brief interview.

The boy spent several months listening to records, tinkering with engines, talking to Dusty. At a later time, when he was asked by an interviewer about his time in jail he looked at his father for approval before responding.

"Not too bad," he said.

When his father quickly interrupted to say what his son *really* meant, the boy fell silent until his father was finished. Then he compromised.

"Well, the food was 'orrible, yeah, but . . . you leave *them* alone, they leave *you* alone."

He talked about his future in very vague terms. "I can't find a job," he said. "I didn't do the killings. I didn't do nowt. But people don't want you around. They don't want to *know* you."

He loved to sit and bury his lips and nose in Dusty's feathers and purr like a cat.

Looking toward his dad for approbation, he added, "In the nick, you ain't got nowt to do but read papers and listen to the radio all day but . . . people *do* leave you alone, don't they. It weren't too bad, really!"

Then he showed his sly little smile and purred, pressing his face into Dusty's ebony wings.

———

Dr. Alec Jeffreys was deluged with media requests. After he went to the FBI's training center in Quantico, Virginia, his face appeared in national newsmagazines in many parts of the world.

His genetic fingerprinting was to be made available commercially in 1987, when the chemical giant ICI started a blood-testing center in Cheshire, England. When reporters asked if he was going to become wealthy as well as famous, the thirty-six-year-old geneticist shrugged off the question, as any good scientist should, but he did not completely avoid publicity. After all, he still wore the "costume," the turtleneck pullover, and he still had the "prop," the hand-rolled Golden Virginia.

If Alec Jeffreys were to become rich as well as famous, few would begrudge him. After all, his discovery was revolutionizing forensic science, as well as providing practical applications in many other areas of science and medicine.

But he knew his life would never be the same after people started recognizing him in shops and pubs, pointing him out as "that genetic-fingerprinting bloke."

———

Eddie and Kath Eastwood had decided to move away from Leicestershire and try for a fresh start in Lincolnshire. They would make only occasional trips back to visit old friends, and to place flowers on Lynda's grave.

But after the inquiry was reopened they were run to earth in Lincolnshire by reporters, and the torment began anew. And Eddie gave a short statement that provided a headline: CATCH HIM AND END THIS HELL.

With the inquiry beginning all over again there was a terrible change in the grief-stricken Ashworth household. Barbara suddenly began experiencing a new emotion: terror.

On dark nights Barbara Ashworth would now almost *see* him at the window, with his red-netted eyeball—the bloodshot eye of madness. In the morning she would actually dread opening the curtains. It became an ordeal. She never knew whether to jerk them open suddenly or try to peer through the opening, with splinters of fear in her stomach.

She had a dream and it recurred. In the dream she was being made to hurry across that footbridge by Ten Pound Lane, the footbridge her daughter had been advised to take, but didn't. A spectral figure with menacing eyes walked behind her, but he was not the man at the window. The face she imagined in waking moments, *that* one was like the face in the Lynda Mann police Identikit: the young man with the spiky punk hairstyle.

Occasionally she had other recurring dreams, of an earlier time when the children were younger and everything was back to normal—the recurring dream of parents of dead children, a dream sure to evoke spasms of grief after waking up. But that dream could be interrupted by the other, by the sound of footsteps behind her on the footbridge.

It finally became impossible to know which was more

fearful, closed curtains at night or opened ones by day. And she was often drawn by day to an upstairs window, the one in Dawn's bedroom, a bedroom kept exactly as Dawn had left it. A room that waited. Barbara could see the churchyard from there. Without willing it, she'd be driven toward that room. She could actually pinpoint the grave, from the bedroom window of her murdered child.

# 18
# Dawn Ashworth II

A peculiar thing happened in the villages of Narborough, Littlethorpe and Enderby. By early December, except for community leaders, people no longer talked about the murders. You would hardly hear them mentioned. When reporters streamed back into the village, now that a new inquiry had been opened, people would shake their heads and move along, refusing to speculate or gossip. They wouldn't even discuss it in the pubs. The tittle-tattle had *ended*.

A spokesman for the Narborough Parish Council probably was correct when he tried to explain his fellow villagers. He said, "People in an English village are more reticent than most, even under ordinary circumstances. And *this* ghastly business, well, they're embarrassed to be talking about murder. It's so devastating it's shut them up."

Three years earlier, some had convinced themselves that the murderer of Lynda Mann might be an outsider, perhaps a transient from the motorway. Now, they could not escape the knowledge that there was a murderer living among them. It was terrifying to parents, and apart from that, it was *shaming* them all. They lived in a civilized place among reasonable folk, and they felt ashamed. So if the footpath phantom was still discussed it was done quietly, in the privacy of one's home.

About the only Narborough villager who *did* talk

openly of it was the village locksmith, the former army
judo instructor who'd been scandalized by seeing the
kitchen porter run his hand up between the legs of the
girl in the Narborough taxi office. The locksmith decided it was high time to do something.

Probably over sixty, his age was belied by a full head
of red hair, a hard muscular physique, and the presence
of a pregnant wife, who, he said, proved that he was
"physically *very* fit."

He was known by villagers to "experiment with alter-
nate lifestyles," this because he was an Englishman
married to an Indian and practiced Buddhism.

It wasn't easy getting accepted by the crusty villagers, the ones in tweed jackets and elbow patches, who
might speak to him one day but not the next. "The
inner village is just that," he said. "It's a *village*, and
you'll never change an English village."

The only people who greeted his family consistently
were the old people whom he described as "infinitely
polite." His Anglo-Indian daughter, born a fortnight after they'd arrived in Narborough, had several "grandparents" in the village who gave her plenty of attention.

Once he made up his mind to take a stand against
terror, he posted a shocking bulletin on the notice
board beside the post office in Narborough village.
Among notices like JUMBLE SALE AT ALL SAINTS CHURCH
HALL and MAY WE INTRODUCE YOU TO CHURCH BELL
RINGING? was a bulletin headlined: PROTECT YOUR DAUGH-
TERS AND YOURSELVES AGAINST RAPE!

As a result, the Narborough Parish Council hall became the scene of self-defense classes on Sunday after-
noons, varying in size from fifteen women to fifty. And
it was the *women* who took part. The adolescent girls,
those who stood to be the next victims of the footpath

killer, would come only when their parents demanded it, and then escape at the first opportunity.

It was frustrating to community leaders that after the kitchen porter was released—and it became certain that the killer was a local man who would strike day or night—the teenage girls, the likely victims, still refused to believe it could happen to *them*. The older ones, the unlikely victims, were convinced *they* were in mortal danger. In fact, the locksmith had to turn away some of the *very* old women to make room for those who at least were possible targets.

No men were allowed to watch the goings-on. Street clothes were permitted and mats were not used, because as the locksmith explained to them, "Rapists don't use mats. And I'm not teaching karate, kung-fu or even judo. I'm teaching defense against *rape*."

Ne'er-do-wells in the pubs snickered about the "roly-poly aunties bouncing around like Bruce Lee, bloodying their noses with their flippin mammaries."

Villagers reported that there was a whiff of chaos and insanity in the air.

Littlethorpe doesn't have the Georgian and Tudor homes found in Narborough. In fact, if it weren't for a large housing estate built in the 1960's it could hardly be called a village. There are two pubs, one of them converted from a big Tudor farmhouse, a few shops, and that's about it, with the exception of Littlethorpe House, a beautiful old Georgian property that dominates the single commercial street. Littlethorpe House was the home of the village squire in earlier times, and cottages adjoining it in the village square were formerly used by the housemaids, the chauffeur and other servants. The present residents of Littlethorpe House are driven about in a Rolls-Royce, a practice that's said to be "too crusty by half," but they lovingly maintain the property.

Down the road from the old part of Littlethorpe village is a housing estate where the majority of the population resides, and it was there that the locksmith went one day, though not to teach self-defense. He did any security work pertaining to the protection of property. The young woman who'd rung him said she might want him to build a protective canopy for their car.

It was a pleasant twenty-minute walk to her home since, as the sign in Narborough village points out, Littlethorpe is just three quarters of a mile down Station Road, past the Narborough depot, across the river Soar, past fields of grass.

When the locksmith arrived, he was met by a polite young woman with a baby in her arms and a three-year-old son hiding behind. She showed him what she had in mind and it seemed a simple enough job. He could always use the money and turned down work only if he sensed the customer wouldn't pay. He didn't sense this at all with Carole Pitchfork.

But then her husband showed up and began asking questions, offering a few opinions.

"He seemed a bit of a know-it-all, a domineering man," the locksmith later told his wife. "I should imagine that inside his own house his wife wouldn't have too much to say."

*That* didn't explain his refusing the work. The man hadn't been rude. He'd only offered a few remarks. The locksmith simply could not articulate the reason for turning down a good job from people who obviously could afford it.

"I just didn't *like* that man" was the only justification

he could offer when he arrived home. The locksmith promised to send the Pitchforks an estimate, but never did.

Modifications were being made in the lifestyle of the villagers. A school bus was rerouted so that no child would have to walk along The Black Pad. The rector of All Saints Church publicly urged young girls to take precautions when going about alone, but of course the girls ignored him. New posters and leaflets were distributed throughout the villages.

After the release of the kitchen porter, the two anonymous donors upped the reward to £20,000 for the killer, bringing journalists running to the new incident room at Wigston Police Station.

Insp. Mick Thomas, who'd been in charge of the house-to-house teams on the Lynda Mann inquiry and had worked on the Dawn Ashworth murder as well, granted a television interview. He was asked if the reward meant that "routine policing had failed and the money was a *bribe,* a sign of desperation."

Looking very professional in a dark business suit, Thomas faced the cameras and said affably, "Oh, no, certainly *not!* The enthusiasm is very good on the enquiry and we're confident we'll get the person responsible."

The enthusiasm was there all right, but so was a note of desperation.

Insp. Derek Pearce was brought into the new murder inquiry to be in charge of the suspect teams, joining Mick Thomas who had the house-to-house teams. Supt. Tony Painter told his two DI's that they were to help him select the very best men available for this one. "We're going to end it once and for all," he said.

As the detectives read Tony Painter, their job was to cement the evidence against the kitchen porter. Police

work is an art, not a science, and an artist is not about
to back down just because he's faced with scientific
hocus-pocus.

Though one of the detectives chosen to see it through
once and for all said, "It seemed to some of us that we
might *have* to accept the genetic fingerprinting, despite
what the gaffer believed. It was either that, or turn to
somebody who *could* detect village murders. Somebody
like Agatha bleedin Christie and her Miss Jane bloody
Marple."

They referred to this new inquiry as "Dawn Ashworth
II." Since the conclusion of "Dawn Ashworth I," culminating with the kitchen porter's arrest back in August,
there had been no work done on the hundreds of messages that had come in. Now with the murder investigation reopened they were deluged once more.

Not having worked on Dawn Ashworth I, Pearce had
to read every message submitted on that inquiry and
examine every lead that had been phoned in since.

As the squad of fifty men began to report and assemble at the incident room in Wigston Police Station, they
were told they were not there to make money. There'd
be no overtime pay and no expenses on this investigation. They were going to work "lates," they were going
to work weekends, they were going to work more hours
than they'd ever worked. They weren't going to be rich
men when *this* one was over.

There really was no confusion as to a starting point. It
was clear to many of the subordinates in the new murder squad that they were not back to square one as the
press believed. Some of their superiors were still obsessed with the kitchen porter, and there was a great
deal of investigative attention directed toward the Carlton Hayes psychiatric hospital where the boy had worked.

Some members of the inquiry who were reassigned

to Dawn Ashworth II would admit privately, "Our job was to get enough evidence to finally make an airtight case against that lad, genetic fingerprinting or no." And so they tried.

On December 18th, *Crimewatch UK* screened a segment on the Narborough murders. It showed the recreation of Dawn's last walk, and a sad little segment with Barbara Ashworth in which she said, "It's not only a daughter that we've lost, but personally I've lost a very good friend. And it's that friendship that I miss more than anything."

All of this was in the hope of persuading anyone who might be shielding the footpath phantom to come forth. The announcer said, "There must be husbands who are suspected by their wives."

The community leaders of Narborough, Littlethorpe and Enderby met on December 6th with Supt. Tony Painter and Chief Supt. David Baker, to offer help and moral support to the new murder inquiry. They'd hoped that all the media from the Midlands would be there but only a reporter from the *Mercury* showed up, causing parish councillors to call the lack of interest "deplorable." The unsolved Narborough murders were getting to be old news and people were losing not just interest but hope.

The next day the *Mercury* ran an editorial that said:

> **The fact is that there are some people within our community who have a suspicion that a member of their family, or a friend, or acquaintance, is the killer or knows who was involved.**
>
> **Knowing now that the whole community has joined the hunt should make the killer and the people foolishly protecting him very uneasy in their beds tonight.**

One of the members of the Narborough Parish Council, solidly behind the police efforts, was the father of Carole Pitchfork of Littlethorpe. She was his only child and he doted on her. It was said he'd never really approved of the young man she'd married.

During the first days of Dawn Ashworth II, Derek Pearce read eighteen hundred messages that had been ignored since the kitchen porter's arrest. After Pearce read them, they were read by Insp. Mick Thomas, after which Pearce read them again.

And after the kitchen porter was discharged by the court there was a message blitz. It took six operators to log the phone calls, sometimes as many as a hundred a day. Messages came from bobbies, from people who'd read newspaper stories or seen television coverage, people who'd had dreams. The operators got messages like "I saw a suspicious bloke in a Birmingham cinema who *must* be your murderer. He laughed when the female star got killed."

Pearce found himself reading new messages about the punk with spiky hair as though the three-year interruption hadn't happened. The orange-haired punk was being sighted all over Leicestershire and hadn't aged a day or changed his hairdo. A punk who, for all they knew, could've gone bald.

A policewoman or other operator would record the new information received by telephone. The messages would go to Pearce who had to spend an enormous number of hours evaluating and deciding what he wanted done. He could give a message a high priority, or put a low code or a medium code on it, or decide it was nothing at all, in which case he'd write, "No further action." The message would then go through the computer system with an action allocator getting the print-out. The action teams might receive a message saying, "Interview subject and eliminate him," or "Trace man seen walking dog in King Edward Avenue at 5 P.M."

As part of the management team Pearce didn't just assign the actions, but made sure the teams followed up on the high-priority markings and didn't just turn to a message that seemed more intriguing. As he put it, he'd "occasionally grab the stickers at the bottom of the piles to make sure the lads weren't sloughing off something promising."

In the pile of eighteen hundred messages that Pearce had to allocate was one that pertained to someone whose name had popped up on the Lynda Mann inquiry because he was unalibied and had a prior arrest record for flashing. Of course, hundreds of names had been called in anonymously by wives, lovers, rivals, neighbors, bosses, employees and nutters, many of those names belonging to people with prior indecency arrests. This one wasn't worth special attention, because in the earlier inquiry he'd been shown *not* to have moved to the village until one month after Lynda Mann's murder.

The anonymous message said: "You ought to have a look at a man in Littlethorpe named Colin Pitchfork."

# 19
# The Blooding

*Blooding*
1. The letting of blood, bleeding; wounding with loss of blood.
2. The action of giving hounds a first taste of and appetite for blood.
                                            *—The Oxford English Dictionary*


By late December, after many members of the inquiry had voluntarily given up their Christmas holiday to work on old and new leads—and after the *Leicester Mercury* had printed a special four-page edition containing every salient fact and photo that might help the police, and shoved this edition into every letter box in the three villages—Supt. Tony Painter and all subordinates were required to suspend their disbelief. It was going to be assumed that genetic fingerprinting actually *worked*.

The ranking officers held a gaffers' meeting with DI's Derek Pearce and Mick Thomas. The subject was *blood*. Chief Supt. David Baker said, "We're going to try something that's never been done."

Baker had sold his superiors on an idea—a campaign of voluntary blood testing for every male resident of the three villages. Anyone who'd been old enough to have murdered Lynda Mann in 1983, young enough to have produced the indications of a strong sperm count found in the Dawn Ashworth semen sample.

Both inspectors felt that Tony Painter was still convinced of the guilt of the kitchen porter. He'd wanted
the Regional Crime Squad to do covert surveillance on
the boy after his release from prison, but the police
administration would not permit it. They knew that
Tony Painter still kept the kitchen porter's file under
lock and key.

But David Baker had apparently begun to believe in
science. Alluding to the kitchen porter, he said that
day, "He's *either* a co-conspirator or he's innocent."

By that they understood that Baker must have been
pondering genetic fingerprinting. Regardless of what
reservations any of them had over the guilt or innocence of the kitchen porter, or the efficacy of genetic
fingerprinting, David Baker had decided that he was
going to seek permission to do it, and he did.

The inspectors privately paraphrased Tony Painter's
reaction as something like: All right, we'll blood-test
them all and then we'll prove that our lad *did* kill Dawn
Ashworth.

In a compromise with his second-in-command, Baker
never admitted publicly that the kitchen porter was
probably innocent. His reasons for blood testing didn't
mention Dawn's killer or killers. He kept it intentionally ambiguous so that everyone could save face.

He simply said to his two DI's, "Find the man who
shed the semen."

The decision was made, and announced the day after
New Year's 1987, that the murder inquiry was about to
embark on a "revolutionary step" in the hunt for the
killer of Lynda Mann and Dawn Ashworth. All unalibied
male residents in the villages between the ages of seventeen and thirty-four years would be asked to submit

blood and saliva samples voluntarily in order to "eliminate them" as suspects in the footpath murders.

The headline on the 2nd of January announced it:

BLOOD TESTS FOR 2,000 IN KILLER HUNT

As several members of the inquiry later said, "We
*had* to have blind faith in genetic fingerprinting."

The planning period for this revolutionary step had
been brief, but the police were publicly assured by the
county council, the parish councils, the rector of
Narborough and the vicar of Enderby that they would
support the scheme. All agreed to urge young men in
the villages to come forward, since no one could be
compelled to do so. One or two community leaders
openly expressed pessimism that the experiment would
succeed, but everyone was running out of other ideas.

The logistical task was far bigger than first antici-
pated. The age span took in anyone who'd been between the ages of fourteen and thirty-one at the time of
the Lynda Mann murder in 1983. Wanted for testing
was every unalibied male who'd worked in or had some
connection with Narborough, Littlethorpe and Enderby,
not merely residents of the three villages. And that
included hundreds of patients at the Carlton Hayes
psychiatric hospital.

In the beginning, testing sessions were set up at two
locations three nights a week from 7:00 to 9:00 P.M.,
and one daytime session was scheduled each week from
9:30 to 11:30 A.M. The young men were asked to come
at a time specified on a form letter sent to each resident
listed on the house-to-house *pro formas*.

Very soon they decided to include every male born
between January 1, 1953, and December 31, 1970, who

"lived, worked, or even had a recreational interest" in the area. A letter was sent to several policemen.

When the Eastwoods were contacted by journalists at their home in Lincolnshire, Eddie said the family was in favor of the tests. He wished they could *force* suspects to take the test, "because the person they want is not going to volunteer."

Of course just about every member of the murder squad believed this as wholeheartedly as did Eddie and Kath Eastwood.

As Derek Pearce later said, "We just hoped that it might somehow flush him out."

By the end of January a thousand men had taken the tests and only a quarter of that number had been cleared. The forensic laboratory was swamped, and it seemed certain that testing was going to take longer than the early estimate of two months. Teams of five doctors were drawing blood at each location, and police reported a 90 percent response to their letters. The 10 percent who did *not* respond were of course the subject of police interest.

Journalists from many parts of the world were now arriving in Leicestershire to try to pry information out of tight-lipped police officials about the unique experiment. Tony Painter relieved his inspectors, Pearce and Thomas, of the television interviews. Painter took them on with the manner of the self-made police administrator. His words were fastidiously chosen, his diction and syntax exact. One reporter writing for an American magazine described his manner as "that of a kindly uncle speaking to a dull child."

True enough, in that up-by-the-bootstraps career cops often think of journalists and *all* civilians as untrustworthy naïfs who can never hope to understand evil and villainy. Painter's style was to let his long upper lip curve into a weary but tantalizing smile, and say, "There are *certain* matters about which I cannot speak." When the kitchen porter had been in custody, everything was "*sub judice*," therefore something about which he could not speak.

He certainly would never reveal who had posted the large reward, causing journalists to waste an hour and a half finding out that they were only a pair of business-men, one of whom employed Barbara Ashworth. The reward hadn't been posted by Boy George or Princess Di. It hadn't even been put up by a local favorite like Humperdinck. It had been posted by public-spirited entrepreneurs who were hardly newsworthy in the first place. If only he'd said that much, but they knew ad-ministrators like Tony Painter were like that. By our secrets we measure our worth.

When journalists asked why police had chosen the youthful age group for blooding, Painter could simply have said, "Rapists are young." Instead, he replied, "There are certain police matters . . ."

But Tony Painter never forgot the Ashworths, and visited them loyally to give encouraging reports of "cer-tain progress."

---

The cops quickly began to refer to their testing ses-sions as bloodings. They would say, "We have to *bloody* this bloke."

The way the bloodings worked was simple. As the donor arrived with his letter, he'd be directed to one of several policemen waiting at a row of tables. He'd be interviewed and identified, no easy task in that there were really no credible identity cards issued in En-gland. The driving license did not bear a photo or a thumbprint, and so the only trusted identity card was a self-employment card, and so-called 714, with the bearer's photo. If he had no current photo the police would take a Polaroid shot which they would later present to a neighbor or employer for identification. Or course, a passport was the best proof of identity.

After being put with a policeman, the subject was asked a few questions (which some resented) having to do with his whereabouts during the two murders. A form was filled out and the donor, along with his form, was taken to a printed register by the same officer who'd conducted the interview. A registration number and an identification sheet were issued, and the donor was walked to a doctor who drew blood and took a saliva sample on a card covered with clinical gauze. A splash of blood was squirted onto another gauze sample card and a sticky label was attached to the syringe, which, minus its disposable needle, became a self-sealing vial.

If the subject was found by the laboratory in Huntingdon not to be a PGM 1 +, A secretor, the analysis generally went no further. If he was, the sample was sent to the government laboratory at Aldermaston where Jeffreys's DNA test was done.

There were action teams, inquiry teams and suspect teams. The miles driven by members of those teams during the next several months were far greater than any ever logged before by the Leicestershire police. These mobile bloodletters swarmed over several counties. They learned to be very clever in suggesting, cajoling, imploring people on their list to come in, but in dozens of cases, a young man simply lived elsewhere and could not comply. In those cases they'd go to him, take him to a doctor's surgery, and authorize payment to the physician for drawing the sample. A team might log as many as five hundred miles in one day on its never-ending quest for blood.

Supt. Tony Painter kept up the pressure from his end. He continued to visit council meetings and church halls and school fetes, and maintained a relationship with all community and church leaders. But it wasn't very easy to get the blood of those who didn't or couldn't respond to the letters. And it was blood they were after now. The murder squad on Dawn Ashworth II was tireless and implacable in its quest. It simply wanted *blood*.

One of the detective constables selected for Dawn Ashworth II was John Dayman. He'd been on Dawn Ashworth I and was among the most popular members of the new murder squad because of his sharp wit and droll sense of humor. He was a good storyteller and impressionist, the kind who would don his cowboy hat and do John Wayne when the urge struck him. DC Dayman was about Pearce's age, a burly cop whose deep baritone voice matched his bulk. He was a tournament-class darts player who seldom had to buy a drink if the bets were on. He was one of those detailed to work on the Carlton Hayes connection.

Since the Lynda Mann murder there'd remained a strong belief that the answer might lie at the psychiatric hospital. On one of Dayman's trips to the hospital, he had occasion to speak with a ward sister about a man the police were interested in blood-testing. During the conversation in her office, Dayman, who felt very uneasy in those surroundings, casually asked what they did if a patient got violent.

The sister was horrified by the question. She said, "There hasn't been a violent patient in . . . I can't *remember* when!" He was chastened by her withering look.

Suddenly there was a scream in the ward outside and a table crashed to the floor. Dayman and the sister ran from her office and found a middle-aged nurse and a male orderly wrestling on the floor with a patient gone berserk.

"He was a ruddy bear!" Dayman later said. "They were trying to get him round the throat and the poor old dear's skirt was up over her head! So I said, 'Do you want any help?' "

But no one answered. Those in the middle of the melee *couldn't,* and the ward sister was paralyzed by the eye-popping spectacle of cellulite pillows bulging from the nurse's white cotton stockings, as the poor old dear groaned pitifully and did Esther Williams–backstroke scissors kicks. The other patients just giggled and drooled, and rooted in their noses or ears, gleefully inspecting the nuggets by crosshatched light from wire mesh windows.

Finally, the orderly wrestling on the floor yelled, "Please! Help!" So Dayman took his glasses off and leaped on the ruddy bear, eventually subduing him. When it was all over, the ward sister recovered her composure, strode up to the exhausted cop and said, "Don't you ever again lay hands on one of our patients!"

It was eerie roaming about the hospital, among people who might be homicidal. Any one of whom might be *him.* Even without rolling on the floor with ruddy bears, John Dayman said it wasn't one of the more desirable assignments: blooding madmen at the nut farm.

Another prime suspect had entered the picture. The murder squad received some tittle-tattle from traveling workers about a certain pipelayer who used to lure young girls into his transit van, to lay some pipe, as it were. The cops got particularly interested when they discovered he'd been working five hundred yards from the Ashworth house.

Then a stunning piece of news energized the entire inquiry. The pipelayer had left work on the very *day* that Dawn disappeared and had gone directly from the job site to a travel agent in Nottingham. When the pipelayer was told he couldn't change a flight he'd already booked, a flight leaving for California in two days, he paid £50 to cancel his ticket. And he coughed up an *additional* £50 to get a new one. It cost him £100 extra to get out of England in a hurry. Furthermore, he never showed up for work to collect £900 in wages owed to him!

The more the police looked into the life of the pipelayer, the more excited they became. They traced him to an old job he'd done in the vicinity of The Black Pad. He'd been laying water pipe in and around Narborough at the time of the Lynda Mann murder!

His ex-wife in Nottingham wouldn't talk to the police about her former husband, and he had no police record for sexual offenses. But he was known to be a violent man, with prior arrests for causing grievous bodily harm. He was a handsome ladies' man, twenty-nine years old, and had worked part-time as a bouncer at a disco.

Tony Painter decided to get in contact with the FBI, hoping that the pipelayer could be located in California and blood-tested. And if that happened, Derek Pearce planned to drive the blood sample straight to the laboratory from Heathrow Airport. Pearce and John Dayman dreamed about going to California to get the sample themselves. Dayman wore an American baseball cap to work for a week, but it didn't help. He didn't even get to go to London with Pearce and Tony Painter to request cooperation from the American embassy.

They never got the blood. Nor did they ever learn why the pipelayer had left in such haste. As with many police investigations the secret ways of people often produced peripheral mysteries as baffling as the one in question.

———————

*He* remained totally inactive for a long time after Dawn. There was only one more to relive sometimes. There was Lynda and Dawn and the hairdresser who sucked it, and there was only one other. She was the first he'd ever touched. It was the first time his exploits made the newspaper, the front page of the Mercury. *It had happened a long time ago, but he could remember it crystal clear. He could remember every moment with all of them crystal clear.*

On February 13, 1979, a seventeen-year-old schoolgirl who lived in the Leicester suburb of Kirby Muxloe was walking home alone by Desford Lane. It was a rather lonely country road and it seemed a bit unusual to see a young man standing there by a farm gateway. But it was 2:30 in the afternoon so she didn't think much about it.

The girl was warmly dressed on that winter day. She wore a brown Shetland wool cardigan, blue jeans and a blanket-style coat. When she got to the gateway and walked past him, his arm whipped out and coiled around her neck.

"Don't scream or I'll kill you," he whispered.

Then he dragged her through the gateway into the open field, the school bag still dangling from her arm. When he got her into the field he pulled her down and knelt on her coat.

"No, no, please!" she sobbed, repeating it over and over.

He grabbed her by the throat and ripped open her blouse. Then he tried to shove his hand down inside her pants, but the zip on the jeans was stuck.

He was very strong—very cold and calculating about everything he did.

Even as she sobbed and pleaded she studied him: pudgy face and meaty lips, full beard and moustache, gingery blond hair. She detected what seemed to be a stain or defect in his front teeth. She saw no mercy in his eyes. She felt utterly powerless.

Suddenly he looked disturbed. He stopped his attack abruptly. He jumped up and ran off without looking back. It was later speculated that, like many rapists, he may have ejaculated before unzipping his trousers.

It was the first. There was such unexpected ecstasy derived from power. From control. From terror. From all the "foreplay."

*When they released the kitchen porter he still wasn't concerned. When they started the blood testing he was. He thought of going to Nottingham to rape and murder once again. Just to divert them from the three villages. The test concerned him. He had faith in science.*

# 20
# The Flasher

The terms sociopath and antisocial personality are often used synonymously with psychopath. Several behavioral signs have been identified by McCord and McCord (1964) and others as characteristic of the psychopath; they are antisocial behavior, impulsivity, hedonism, aggressiveness, guiltlessness, a warped capacity for love, and the ability to appear superficially adequate.
—DAVID C. RIMM and JOHN W. SOMERVILL,
*Abnormal Psychology*


In the summer of 1979, when Carole was only eighteen, she finished her course in preliminary social work and took an appointment as a volunteer at Dr. Barnardo's Children's Home in Leicester. A charitable organization, Dr. Barnardo's had been founded in the 19th century as an orphans' asylum, but in modern times had become a residential home for mentally handicapped children. Carole was an "auntie," a residential social worker. When Colin Pitchfork came to Dr. Barnardo's as a volunteer he was nineteen, just eight months older than she, but Colin had already been employed by Hampshires Bakery for three years.

Carole was always given the hard jobs at the home, like washing and ironing, but Colin seemed to get the pleasant assignments, like playing with the children and organizing activities, and baking cakes for them.

A pudding of a girl, plump and approachable, Carole

233

had bird's-egg-blue eyes that narrowed to slits when she laughed. Colin, on the other hand, was quiet spoken, but he had a cynical arch to his brows, as though he were repressing an urge to sneer. A ginger blond, bulky through the trunk and shoulders, Colin sometimes sported a beard, sometimes not. He had a bit of a saw-toothed grin, but if he kept his mouth closed he was very presentable.

They seemed quite different, Carole and Colin, but being the same age, they drifted toward one another during those days at Dr. Barnardo's. She said that after her parents divorced, Colin "was a pillar to lean on." A very shy chap until you got to know him, Carole told her friends.

In August of that year Colin asked her to go out for a drink. He wasn't a pub man and didn't drink often, but when he did, he talked. Carole learned that Colin was a local lad from Leicester, the second child in a family she came to call matriarchal. Colin's older sister was quite assertive like her mother, and was bent on being a doctor with an ultimate goal of research in microbiology or biochemistry. His younger brother was intent on taking a degree in engineering. Colin, the middle child, told Carole that he was "the black sheep" and the "underdog."

*We lived in the village of Newbold Verdon then, me mum and dad, a sister and brother. The older sister were very clever, always Mummy's little girl. They paid for her all the way through her education clear up till when she was twenty-seven years old. And the younger brother, he were the baby so they doted on him. I was in the middle, but I was never neglected. I joined the Scouts because Mum expected it. She was heavy into the Scout movement. They were proud of me then.*

*But I remember how the other boys used to make fun of me in school when I went in the shower because I were bigger than most and had pubic hair before the others. I guess the problems started at home. I used to show it to girls I knew. Right there at my own house from the age of eleven. I used to like to show it. I started going out on the streets to do it. I started showing it to strange girls.*

*Then came the good part of life. I went to Norway one time with the Scouts, and when the Scout leader left the group, I became a Scout leader with Mum's help. I became a hero at home then because I was doing so good. Everyone found it exceptional to be a Scout leader at the age of fifteen. The good side of life was always good and the bad side was bad and there seemed to be no in-betweens. That lasted for several months until a new Scout leader got appointed and then all the praise stopped.*

*I enjoyed English, but for me the majority of subjects was a waste of time. And the same with church. Me mother was religious but I found it all hypocritical. But I attended church and I became a server, which I hated. All the bloody church ever done was make money from old ladies to make itself rich.*

*Before I gave up on school at sixteen, I made a film on vandalism and it got shown at some local high schools. I loved the glory I got from that, but school weren't for me. The thing I most hated was everyone calling me "her brother," not feeling like a person in me own right. But I was still a Scout then, at least.*

There were things he told Carole during the time of their courtship, and things she discovered later. He told her a bit about his history of flashing. He even implied that it used to give him some of his greatest thrills. She tried not to worry about it. She was in love with him.

*The first time I got caught flashing it had an effect on me mother of horror and upset. I got visited by a police constable who gave me a lecture and told me how to avoid a similar incident. That was worth bloody nowt. Then I got caught doing another and had to go to court and it got reported in the local paper and they kicked me out of the Scout movement. It was bad because me mum was the group Scout leader in Newbold. But they allowed me to join the Glenfield Scout Group.*

Colin told Carole that he'd wanted to try for a Queen's Scout Award and even the Duke of Edinburgh's Gold Scout Award because his mother would've liked it. As part of the Duke of Edinburgh's award he'd had to do voluntary service, which brought him in contact with Dr. Barnardo's Home.

*I worked there for five years, taking care of those mentally handicapped kids, taking them for days out. The female members of the staff looked up to me. Me mum figured I was back on the straight and narrow and I never did no flashing at Glenfield because of my work at Dr. Barnardo's.*

*But you get that need. You go out sometimes and cover fifty or sixty miles looking for that opportunity. It's the high I needed. Yet sometimes I didn't get nothing out of it. You never knew how it would turn out. Then they caught me again.*

After her parents' divorce was final, Carole moved into Dr. Barnardo's and saw more and more of Colin Pitchfork. They eventually became engaged and lived as husband and wife, but during the engagement, even after they'd set the date, it happened *again*. Colin was summoned before the court for indecently exposing himself to young girls.

"I was naïve," Carole later explained. "I simply didn't understand the business of flashing. I actually thought it was like giving up smoking or going on a diet. That it'd be difficult for a few weeks and then just go away. I had no idea how complex it all is. Agoraphobia is about the only abnormality that I can understand perfectly. Any other abnormal behavior, I can't. Anyway, according to Colin, the probation service told him he'd *outgrow* his problem, even though he was twenty-one years old at the time. They put him on probation again."

In May of 1981, they were married in grand style. Lace gown and veil, top hat and gloves—the young couple did themselves proud.

Colin and Carole settled in Barclay Street in Leicester, and Colin continued working at Hampshires Bakery. He had an artistic side, a flair for drawing, and he was keen on learning to decorate cakes. He was also musical, having played tuba as a boy, and would sit for hours at a keyboard, a piano or a steel drum.

Carole didn't just love her young husband, she admired him. But it took awhile for her to understand Colin Pitchfork, and to sort out the relationships in his family. For instance, Colin didn't know much about his parents, not even how they'd met. He knew only that his father had formerly been in the mines in Chesterfield, and that his mother had been raised in the house she'd raised Colin in, a house his parents retained after his maternal grandparents died.

"They just didn't *talk* about things" was how he explained his family to Carole.

She, on the other hand, knew *all* about her family. She was a daddy's girl and always had been. Hers had been the kind of family where they'd kiss each other as

quickly as saying hello. And she'd had her own horses
as a child, her last being a show jumper named Jamie.
She'd ridden in competition since she was seven years
old and her dad was always there to watch her win
ribbons. He was a civil engineer who sold sewer work
and motorway construction to government clients. He'd
always worked in a jacket and tie at a middle-class job.

After Carole's father remarried, he and his new wife
moved to Narborough village where he became one of
the sixteen-member parish council. He was an active
youngish man who loved boats, shooting and his daughter. His disapproval of his son-in-law, Colin Pitchfork,
was even greater after Colin was caught flashing yet
*again*, and was sent for psychiatric counseling to Carlton Hayes Hospital in Narborough. To The Woodlands,
there by The Black Pad.

*I got dealt with by a probation officer and a doctor. I*
*were referred to The Woodlands because that's where*
*they take outpatients. A waste of time. A bleedin waste.*
*Probation officers and psychiatrists, these people are*
*quite happy if you tell them what they want to hear. I*
*can look at the two sides of my life so objectively. I look*
*on meself as quite intelligent, and I can't believe how*
*easy it is to spin yarns to these people.*

*That particular flashing arrest never got in the pa-*
*pers, so life carried on, really. And the flashing were*
*also carrying on at a nice little rate. I could see it*
*getting up into the thousands in twenty years.*

Colin seemed very placid to Carole during the months
of her pregnancy in the spring of 1983. He seemed to
enjoy a class he was taking in cake decorating at
Southfields College. It was an evening class, and he was

often tired from working at the bakery, but he was still anxious to get to the community college.

Soon he was riding to school with another student who would drop by and pick him up. Leslie was eighteen years old. To her he seemed worldly and experienced. He certainly knew a lot about making cakes and decorating them.

Colin began ringing Leslie at her home, and then seeing her secretly. Soon, when she came to pick him up at his home he'd invite her in for tea.

Carole had a part-time job working with children at Venture Playground. One day when she was at work, Leslie and Colin had sex in Carole's bed. It seemed foolhardy to Leslie, but that's what he wanted. He used a contraceptive and the sex was straight-forward. She later said he was kind to her. She'd been a virgin up until then.

Leslie was attractive, a fact not lost on Carole Pitchfork. And she was young. Perhaps she had a way of looking at Colin when she'd come to pick him up for class. Perhaps there was the scent of another woman in Carole's bed.

"Why do I keep putting two and two together and getting five?" Carole asked her husband one night, after she'd accused him of being a bit too chummy with the girl.

"I just went out to the pub for a drink!" Colin told her. "One drink!"

"You don't have to go so often."

"Twice a week is too bleedin often? Am I a drunkard too?"

He wasn't a drinking man, nobody's pub mate, and since Carole was pregnant she really didn't feel like sitting with him in a smoky pub when she could be at home.

"I've asked you to come along, haven't I?" he said. And that was true.

But one evening, after her fourth or fifth accusation, after he was just a bit *too* late coming home from the community college, he simply said, "All right, it's true."

He showed no remorse or sorrow. It was a simple fact and it was pointless to deny it. It turned out he'd sent flowers to Leslie, which may have been his undoing.

Carole, of course, was heartbroken, and tearfully vowed to leave him forever.

When she moved out and returned to her mother's house for two weeks, her dad learned where Leslie lived. He went directly to the girl's home to inform her father that his teenage daughter was having an affair with a married man. The evidence of flowers brought forth a confession from Leslie, and she agreed to break it off for good.

*The flashing stopped the whole time I were seeing Leslie. It stopped because I got a lot of excitement from her, combined with the excitement that Carole would catch me.*

The Narborough parish councillor was not a man to let Colin Pitchfork trifle with his daughter. In fact, he would've been totally in favor of a divorce, but his son-in-law was a persuasive talker. Persuasive with Carole.

She agreed to meet with him, and he promised he wouldn't do it again, and when he was all finished explaining, *she* was the one who felt guilty. Why *hadn't* she gone out with him to a pub on those nights when he'd asked? Didn't she know a man needed companionship when things were going wrong? Wasn't he a good provider? Was he ever unkind to her? Did he ever so much as shout at her in anger? Wasn't *she* the one who'd done all the shouting during their rows? That girl Leslie was just . . . *convenient*.

"Just one of them things," he said to her.

"When you love somebody it's easy to blame yourself," Carole later explained. "Even when you feel such a prize fool for doing it."

"Leslie was like the flashing," he explained to Carole. "Something I got a buzz from because it was something I shouldn't do."

"He said she gave him something I couldn't give because she was so *young*," Carole Pitchfork later related, but Carole herself was not yet twenty-three years old in the summer of 1983, and the comparison made her steam.

"What the bloody hell do you mean?" she said to Colin. "I'm too *old*, am I?"

"No," he said in his quiet, reasonable way. "But Leslie, she was a *virgin*. And you weren't, were you?"

It was the first time he'd ever said anything like *that* to her. She'd never forget it, but she'd forgive him almost anything. With Carole, outrage could always be diminished by guilt.

"It was just one of them things, love," he repeated, when he had her emotionally subdued. "Our child will need *two* parents. And if you don't give people a second chance you'll always wonder if you *ought* to've done, won't you?"

Even before the birth of their baby, Carole had a yearning to get away from that house in Leicester.

"I want a garden," she told her husband. "And we need a different environment for our child."

Privately, she told friends that she wanted to get away from "bad memories attached to the house." Presumably, memories of Colin and Leslie alone in her bed when she was dragging her pregnant belly around a playground, working with other people's kids while dreaming of her own.

When their son was born in August, Colin was moved to tears and those tears helped wash away her lingering bitterness. She decided to turn the page and try to forget, but it was all becoming just a bit *too* much, like those romances she read and quickly forgot.

With her father on the Narborough Parish Council, the village was a logical destination. Just down Station Road from Narborough was a nice housing estate with several homes for sale. She and Colin had a look and settled on a semi-detached two-story house in a cul-de-sac with a whimsical name: Haybarn Close. It seemed propitious: Pitchforks in a Haybarn.

Carole planned the move to Littlethorpe in December. She wanted to have Christmas there in the village, and she desperately wanted what most want at a later time in life—a new beginning.

Of his family members, Colin's father was closest to him. When the older man had time off from his job as a steelworker he'd visit and offer his labor. He'd helped Colin build a new fireplace on one of those visits.

Whenever he came, Carole only had to say something like "I have a casserole on for dinner." And he'd eagerly say, "Oh, I haven't had one in years!"

"I think he comes to get away from home," Carole told her friends. "My mother-in-law has a heart of gold, *but* . . ."

Colin's parents were more excited about having a grandchild than Carole's were. "My parents had their own separate lives," she later explained, "but Colin's were child oriented. They wanted to take care of the baby whenever they could get him. Still, they're not very approachable people, except for Colin's younger brother. His brother became closer to me than he was with his own sister. He deals with the rest of his family by ignoring them."

Well, she hadn't married his family, she reminded herself, and he hadn't married hers. Carole was simply grateful that after the baby was born Colin and her father seemed to be getting on a *bit* better.

But if Colin and Carole Pitchfork had little in common with their in-laws, they were finding more in common with each other. They both enjoyed the outdoors, and looked forward to village life where they could indulge their passion for walking. And anyone who'd lived in the villages knew there were many lovely footpaths in Narborough, Littlethorpe and Enderby.

Carole had signed up for a night class at Rowley Fields College in the fall of 1983. Colin wasn't all that keen to see her "getting education," but he didn't object. Before leaving their old home in Leicester for the new one in Littlethorpe, she thought they should have a "leaving party" for a few friends and neighbors. They planned it for late November, a month before the move.

Five days before the Saturday night "do," Colin decided to make tapes from several records, so they'd have continuous music for the party. He began the record taping while she was getting ready for her Monday night class.

"I'll have them done by the time you get home," he promised her.

It was just before 7:00 P.M. when he drove Carole to school in their Ford Escort. The baby slept in a carrycot on the backseat.

"See you at nine," he said when he left her.

\* \* \*

Colin Pitchfork collected his wife at nine o'clock. When they got home she was happy to see the taping was nearly finished. She was always uneasy leaving him alone at night. He still had a roving eye—she didn't fool herself about that.

It was good to get home on that clear frosty night. The baby looked snug enough, but Carole was freezing. Monday, November 21, 1983, was the coldest night of the year.

# 21
# Phantom Days

ANTISOCIAL REACTIONS (*psychopathic personality*)
. . . Superficially the sociopath excels in social situations. . . . On the other hand, those who become better acquainted are soon aware of his immaturity, superficiality, and chronic inability to make a success of his own life. . . . He becomes a true disappointment to those who were charmed to expect more, to those who began to believe in him, to those who continue to see his potential, to those who still hope for him.
—RICHARD M. SUINN, *Fundamentals of Behavior Pathology*

While Carole was pregnant, Colin became "unsettled" again, and he started complaining more. He'd worked at the bakery since he was sixteen years old and he was tired of the grind, having to be at work at 5:00 A.M. He ought to try other things, he said.

"Perhaps I'll try writing," he announced.

"First you should try reading a book or two," she answered.

Carole read a book a day sometimes, the kind you forget an hour later, but at least she read. Colin didn't read except on holiday, perhaps one book a year. He looked at the newspaper and *Reader's Digest* and that was all.

"Perhaps I'll just start looking round for another job,"
he said from time to time.

"The grass was always greener for him," Carole recalled. "When we had a son, he wanted a girl. His job
was too boring. Our house was too small. There was
*always* something wrong."

At the time, Carole Pitchfork wasn't aware that the
restlessness she saw was prompting more romantic en-
tanglements, more *risks*. Life in Littlethorpe was happy
enough for her. Her father and husband were still not
friends, but at least they got along, now that Colin was
appearing more domestic. Even through the terrible
twos when their son was, like all children his age, a fair
handful, Colin was ever patient. He never spanked the
child or shouted at him. Carole told her friends that she
could yell her head off to no avail, but Colin had only to
raise his voice an octave and the tot would pay attention. Colin Pitchfork wanted a girl, but their second
child, born in January of 1986, was another boy.

That spring things were back to normal in the villages
as far as the footpath killing was concerned. Most seemed
satisfied that The Black Pad murder had been an
aberration—an explicable tragedy that happens and is
never repeated. The killer must have been passing
through, perhaps on his way to London. He could've
stopped *anywhere* and done his terrible deed.

Carole recalled that when the constable had come to
their home on the house-to-house check during the
Lynda Mann inquiry back in 1983, Colin had been
quite willing to answer the policeman's questions even
though there was some concern about his past record
for flashing. But the Pitchforks were never bothered
again after that first visit, so they seldom talked about
The Black Pad affair. Only once, that she could remember.

Commenting on that time in Leicestershire when the county seemed to be a repository for bodies, Colin said to Carole, "This seems an ideal place to commit murder."

When she asked why, he said, "Because there's so *many* unsolved murders being investigated."

A seventeen-year-old girl who worked at the Queens Road shop of Hampshires Bakery met Colin Pitchfork when he brought baked goods in the company van. He seemed to her like a decent sort of chap, a good family type who always talked about a son he adored. He was so quiet spoken that she had trouble understanding him. Sometimes it was necessary to move closer.

One day in November she was washing up in the kitchen of the bakery when she was startled by hot breath on her neck. She whirled to find him there, grinning. She tried to shove him away, but he didn't budge. Then he put his hands on the boiler, on either side of her, and stared. This time she shoved past him and went back to work. He laughed and made light of it the next time he made a delivery.

In May of 1986, she was about to celebrate her eighteenth birthday and several employees were invited to a party in her honor. Colin Pitchfork was given the job of baking and decorating the cake for the do, and stopped by her home to make arrangements. He seemed *very* nice and she thought perhaps she'd misjudged him. He invited her to a pub for a drink one evening and she went.

When they left the pub he drove her home but missed the road to her house.

"I took the wrong turning," he said, but after a few minutes he took *another* wrong turning and she found herself on a lonely country lane.

He stopped the car abruptly, leaned over and kissed her lightly on the cheek.

"No! Take me home!" she told him.

He complied meekly, drove her home and said good night.

Except on her job, she didn't see him anymore. But he *did* leave poems with her. There were five altogether. In one of the "love poems" he expressed a fervent desire to impregnate her with his child.

She failed to respond to the invitation and her lack of interest in a creative act to replicate part of himself may have annoyed him enormously.

The love poems stopped, and the poet returned to the prospect of a dreary life in the bakery with a remote possibility of being a foreman someday.

The supervisor of Colin Pitchfork at Hampshires Bakery had this to say about his subordinate: "He was a good worker and timekeeper, but he was moody. Almost a barrack-room lawyer at times. And he couldn't leave women employees alone. He was always chatting them up."

Colin had been employed at Hampshires since August of 1976 when he'd started as an apprentice baker and confectioner. During all those years his barrack-room lawyering had gotten him sacked several times. But somehow he'd always manage to talk his way back into the gaffer's good graces before the dismissal order came into effect.

Along with Colin's boss, Carole must've had a notion about Colin's attention to female employees. She was always ringing the bakery on some pretext or another to check on him, to make sure he was there.

He'd had to attend technical colleges as part of his training, to get his apprentice papers, but he'd despised the classroom part of it. What Colin liked about his job was the artistry, the decorating of cakes. Colin's photograph appeared in the newspaper because of one of the cakes he'd done. It was an impressive job, beautifully composed. He'd sculpted a night rider's motorcycle with colorful icings. But he was asked to pose beside the cake for a publicity photograph.

That was an extraordinary photo session. The cake's creator looked about as sanguine as a hostage in Beirut. In what should have been a happy news photo he seemed to be staring warily at the lens, at some terrible threat that a newspaper photo might spawn.

In July, 1986, Colin became depressed, for no reason Carole could fathom. He seemed always tired but couldn't get to sleep. He went to a medical center and was given some sleeping tablets. Carole didn't find out until a later time what was on his mind: his lover's pregnancy.

The affair with the woman he called Brown Eyes hurt Carole most of all. She simply couldn't begin to comprehend *that* one when it came to light. Leslie had been young and pretty, but this one? It seemed impossible.

Brown Eyes worked as a baker in the Hampshires shop on King Richard II Road. She had a baby of her own and was in the process of getting a divorce, her husband having left one month earlier. According to the gossip Carole got from Colin, the tall young woman had once thrown boiling jam on a fellow employee. But there must've been an unspoken message received or imagined by Colin Pitchfork.

One evening he showed up at her home and simply said to her, "Everyone at work has a bet that I daren't come round for a cup of tea."

"And you were cheeky enough to do it."

"At least I'm honest," he said, and she invited him inside.

He had his tea and they chatted for fifteen minutes. Then he was off, heading for home on his moped.

But he came back, regularly, and always uninvited. Brown Eyes didn't turn him away. She needed him to commiserate about her failed marriage and found him to be what she called "a sympathetic friend."

Soon the others at work started noticing. Colin was too solicitous and attentive to her. He started lifting heavy things off the shelves so she wouldn't have to.

On a cold day in December she was wearing knee socks and he told her, "I *like* white knee socks."

For a month he came and went from her house, kissing her only on the cheek. But one day in December things took a turn.

"The sex began," she later said. "It happened about once a week, on weekdays only. There was never any real foreplay involved. Just straightforward sex."

As with Leslie, Colin invited Brown Eyes to his home and introduced her as "a friend from work." She brought her infant along and even babysat for the Pitchforks.

After Carole had her second child in January, Brown Eyes continued to visit, and when Carole went back to work, Colin and Brown Eyes would often be there at the house decorating a cake or playing with all three children.

They were just the best of friends, Colin said to Carole, who told herself that this woman wasn't his type. She wasn't innocent. She wasn't young. She wasn't a virgin.

The happy arrangement teetered when Brown Eyes announced to Colin that she was pregnant, but Colin responded in his measured quiet way. He overwhelmed her by saying he was *pleased*. Colin advised her to keep the baby or abort it, as she saw fit, and if she chose an abortion he'd gladly pay for it. He seemed concerned, and visited her often to discuss the decision.

She made an appointment with an abortion clinic, but at the last minute canceled. She decided to have Colin Pitchfork's baby, come what may.

When she told him of her final decision, he said, "Good. Now see a doctor and take care of yourself. I'm happy. I've always wanted a girl."

———

During the late summer and fall of 1986, when the shocking murder on Ten Pound Lane had the villages in the grip of terror once again, notices were posted of a special bus pickup to Brockington School. The pickup was in Littlethorpe, by the Pitchfork home.

As terrible a tragedy as it was, the footpath murder inquiry did not seem to be an event about which a young mother like Carole Pitchfork should fret unduly, as long as she was prudent where she walked. And most of the time now, she drove the car wherever she needed to go.

Colin didn't watch television much, having to get up so early to work at the bakery. Yet he watched *Crimewatch UK* and all the news bulletins about the Ashworth inquiry, especially after the release of the kitchen porter, when the villagers were distressed and confused.

Carole arrived home from work ahead of schedule on the night they announced the kitchen porter's release. She liked to pop home unexpectedly, just in case.

"Bunked off early," she said, finding him glued to the television.

"They let him go," he said.

"If he wasn't the one, now what?" she said. "I don't think they'll catch *anyone* until it happens again."

He shook his head and said, "They just haven't a clue, have they?"

"Colin wouldn't balk at having a go at something," Carole said. "He always wanted to try racetrack rallying but it cost sixty quid a day and we couldn't afford it. But if we could've, he wouldn't have balked at having a go."

Whatever it was he cared to have a go at, you could be sure it was something for him, whether or not she was interested or included.

"The life of the family had to revolve around Colin," she recalled. "And I had to be at his beck and call. Even when I started asserting myself by going to college."

Carole had some ideas about getting a proper education now, but whenever she'd get excited about a college class schedule and try to share it with him, he'd listen a moment or look at her material and say, "Looks like a load of bullshit to me."

"You use other people's egos to step on just so you can improve your own," she told him. "You pull me down to build yourself up."

"I ain't a college man with fancy theories. So leave off."

"There you go," Carole told him. "What you're really doing is comparing yourself to your brother and sister. Just as you've always done."

"You're too clever by half," he said. "Know everything, but know bugger all!"

The young mother was yearning to learn and grow, and learning about her husband was to be a large part of her curriculum. Yet, in retrospect, she couldn't really say that he'd ever tried to learn anything about *her*.

"Colin's like an adrenaline addict," Carole told her closest friends. "Always needs something to psych him up. He can talk and talk, and I have to listen. Bores me silly with details about *his* schemes, but we never talk about *me*. Never about my work."

As to his unstated feelings about Carole's job, she later reflected on them by saying, "My work must've threatened him because it was a world he had no control over. I was the boss there. People would do as I say. I wanted to take a course in certificate qualification for social work, in order to become a probation officer. He claimed we couldn't afford it, but I came to realize it was because he was afraid he'd lose his grip over me. *Another* person in his family with an education. He couldn't bear that."

Of Colin's easygoing, quiet-spoken ways, Carole said, "He never raised a hand in our house no matter how much I yelled at him, but make no mistake, he was in control. To an outsider it might not appear so, but Colin Pitchfork couldn't have lived with someone without being totally in control."

Worrying about his prematurely receding hairline and expanding waistline, about getting bald and going to fat, Colin played squash at the Leisure Centre in Enderby and sometimes attended the Littlethorpe Judo Club, but his attempt at more strenuous athletics ended when he hurt his knee playing rugby in October, 1986. He was on medical leave for eight weeks, and they celebrated Carole's twenty-sixth birthday in November while he recovered. His leg wasn't much good until the end of the year.

The inactivity was making Colin unsettled again. He'd always wanted to set up a business based upon cake decorating. Not ordinary cakes like those he baked at

work, but exquisite cakes, made to order. Customers
would commission the designs and he'd sell accessories
to go with his own creations. They'd be very expensive,
but the best cakes anywhere.

Carole Pitchfork had heard variations on that theme
since they'd been married. She'd heard dozens of
schemes and watched him write down figures, and com-
pute loan payments, and add up the interest they'd
have to pay on a loan.

She finally said, "Look, I'm sick to death of all that
scribbling on bits of paper. Get on with it!"

And then she felt guilty for not being more supportive and quickly offered encouragement. "You can do
almost anything you set out to do! You're bright. Brighter
than me. Your mind turns to almost anything. Just *do*
it. I'll back you up, whatever you decide."

But Carole had serious doubts he'd ever leave his job
to try the scheme. In the first place, he'd be afraid it
wouldn't be the best cake-design studio in all of Britain.

"He had to live up to certain expectations," she said.
"He had to think he'd achieve greater success than his
sister and brother, or he'd never make a move as far as
a career's concerned."

"I'll have a real cake-design studio," he promised.
"You'll see."

He even had a name for it. Their older child's middle
name was Ashley and the baby's was James. The studio
would be called James Ashley's.

He used a few of those unsettled days to paint a
charming mural on the wall of the children's bedroom.
It was a colorful cartoon of Thomas the Tank, the blue
train engine with enormous eyes and a huge smile that
whistles through the children's storybook on various
adventures. The mural was six feet by twelve feet,
skillfully executed. In it, Thomas was steaming into

Narborough Station, chugging into the dreams of the sleeping village tots. Carole was proud of that mural.

Yet the everyday life Colin led seemed to him demeaning and bogus. And once he remarked wistfully to Carole that marital sex was not as exciting as the other kind. But he quickly admitted that the other kind would destroy their marriage. And anyway, he was usually too tired for sex. Carole had to take the initiative most of the time. There *was* a way to arouse him. He liked her to wear long white socks. The kind a schoolgirl wears.

# 22
# The Test

Connected to the antisocial personality's moral insensitivity is his lack of any feelings beyond the superficial. . . . The antisocial personality is typically cynical, ungrateful, disloyal and exploitative. He has no empathy or fellow feeling and therefore cannot comprehend on an emotional level how his actions hurt others. Other people are there to be used. As for giving or receiving love, these are beyond his capabilities. As a consequence of his lack of strong feelings the sex life of the antisocial personality is typically manipulative and faithless.

—JAMES F. CALHOUN, *Abnormal Psychology: Current Perspectives*

For Colin Pitchfork, January, 1987, had to have been the worst month of his life. Brown Eyes had worked at the bakery until the last week of that month when she started getting severe pain in her legs. The pain got so bad that one day she couldn't get out of bed to go to her job. Later that afternoon, the twenty-nine-year-old woman delivered her own baby.

"I pulled it out, and put it up to my chest," she told authorities. "I saw it wasn't breathing and I tried to give it mouth-to-mouth. Then I rang a doctor and my parents."

261

She was taken to hospital by ambulance. At 7:45 P.M. that night, Colin Pitchfork arrived at her bedside in tears, asking where their baby's body was. When he was told it was at the Leicester Royal Infirmary, he said he wanted to go there and view the remains of his daughter.

In fact, he wanted to do what Robin Ashworth had been forced to do five months earlier.

There was another reason why January had to have been the worst month in the life of Colin Pitchfork. He'd received a letter from the police about the voluntary blood testing, and was given a date and time to report. He told Carole he was afraid to give blood.

"Why?" Carole demanded.

"The flashing!" he said. "Don't you know the coppers're going to take one look at that flashing record and give me trouble?"

"That was a long time ago," she said. "That won't be a problem."

"You don't know them," he said. "They're looking for somebody. Anybody."

"Well then, give them the blood and they'll see they're not looking for *you*, won't they?"

"Meantime they'll give me all *sorts* of bloody hell," he grumbled.

When the second letter came two weeks later and he *again* failed to report, Carole started getting uneasy.

"This no longer makes any sense, Colin," she said. "You *are* going to take that test. Now *I'm* the one demanding it, not the bleedin coppers!"

"And if I get chucked in the nick for me past record while they waste time sorting out the test, will *that* satisfy you?"

"Take the test," she said.

Carole Pitchfork's anxiety was unfocused. If there was ever a fleeting specter of recognition it remained too fearsome to face.

In late January, a twenty-five-year-old employee at Hampshires Bakery was approached by Colin Pitchfork and asked to accompany him to the loading bay. When they were alone Colin told a strange story about having been arrested for flashing when he was very young. Colin talked about the massive blood testing going on in his village and said he was afraid to give a blood sample because of his admittedly unreasonable terror of police. He asked his fellow worker if he'd consider giving it *for* him. Colin said how easy it would be to insert his friend's photo into Colin's passport. He offered the coworker £200 for his trouble.

The baker turned him down at once, saying that Colin should forget the past.

Colin said, "*Think* about it."

The same day, Colin approached him again and said, "You could do it easy. There's no way it can be traced."

But the baker again refused him, advising Colin that his fear of policemen would be dissipated by confronting it.

Colin had another acquaintance and fellow baker at Hampshires who'd offered him a bed for a few days if he ever again got kicked out by Carole. Colin told this baker a story of how he'd been arrested as a younger man for a flashing offense after he'd been unjustly accused by some hysterical females who'd seen him urinating.

"I were innocent," he said, "but that didn't matter none, and it wouldn't matter now. I could be set up because of me past record!"

"It's too serious a charge to be setting you up," the baker assured him.

"Well they might make a mistake on the test, mightn't they?" Colin argued. "I'm scared to go up there."

"If they *did* set you up, a solicitor would tell you to plead not guilty and you could demand a new test," the baker said. "You got nothing to fear."

"I can agree with what you say *in principle*," Colin said, "but I'm still too bleedin scared of coppers to go."

"I wouldn't be afraid if I was you."

"Then do it *for* me!" Colin urged. "I can't offer you no more than fifty quid but you're welcome to it. You could use my passport. I can put your photo inside the embossed stamp. It's a minor trick. No problem at all."

"Your fear just ain't rational," the baker said. "I'll go with you to the test. You'll have me right there with you for moral support."

Yet another baker, an eighteen-year-old, was approached by Colin Pitchfork that same week and was told the same flashing story. But this young man told Colin it wasn't right and he absolutely wouldn't do it. Colin never bothered him again.

A "charge hand baker" for Hampshires had occasion to talk to Colin that week. Colin went to the older man's home and wept yet again over the death of the baby. Colin told the baker he was going to the mortuary to take a photo of the infant. He also gave the older man the four-page murder supplement that the *Leicester Mercury* had delivered to every home in the three villages. The baker had expressed interest in reading about the murders.

The death of his daughter was simply breaking his heart, Colin told him between sobs. He wasn't in the mood to chat about murder. The river Soar was dryer than Colin Pitchfork.

Ian Kelly was quick to say, "I'm English, not Irish. All the Irish Kelleys have two *e*'s in the surname."

Not true, but Ian had heard it somewhere and accepted it on faith. He was like that, accepting things on faith.

"He's very easily led," his young wife explained in his presence.

He was twenty-four years old, and worked at Hampshires as an oven man. He knew Colin as second in command to the foreman. Ian had never socialized with Colin Pitchfork away from the job but they got on well enough at work.

Ian had worked at Hampshires for only half a year. He'd gone to community college for three years to learn baking and confectionary, and admired the work Colin could do, especially the night rider on the motorcycle. He knew Colin had a special flair with cake decorating that few could match.

Ian had a shy, sincere way of talking and he was appealing in a pallid, wispy sort of way. He looked as though he could use protection, and his twenty-one-year-old wife, Susan, was probably the right woman for the job. Of Spanish parentage, she was sturdy, assertive and generous. She could usually take care of herself and Ian as well.

Susan Kelly had first encountered Colin Pitchfork at the bakery Christmas party. He'd asked her to dance and everyone was having drinks and a good time so she accepted. While they were dancing, a bit too closely to suit her, his hand dropped a few inches down from the small of her back.

He pulled back, grinned, and said, "Kelly's come up

trumps getting a girl like you for a wife. How'd you like to go outside?"

"For what?" she asked.

"For a fuck," he answered.

He didn't seem to be *that* drunk. The fiery young wife of Ian Kelly said to Colin, "Unless you want a sharp kick between the knees you better watch your mouth. *And* your bloody hands!" With that she whirled and stormed back to their table.

All she'd said to her husband was "I don't like dancing with him. The way he leers just gives me the creeps." And *that* was true enough.

She later wished she'd told Ian about it instead of leaving it. Perhaps if she had, he'd never have agreed to help his co-worker with the "little spot of bother" that had come up in his life.

"I been sent a letter about the blood test," Colin told Ian at work, in the presence of another employee.

It wasn't the first time he'd mentioned it, and they were all aware of the Narborough murders and the mass blood testing.

"So take the test," the other employee said.

"I'm scared to go up," Colin said. "Scared to death of cops."

"I'll go up with you," the other man offered. "Only a bunch of coppers and a little needle, isn't it?"

Colin dropped it, but a few days later, he took Ian Kelly aside and said, "I *can't* take the test!"

"Why?" Ian asked.

"I already took it! See, this *other* bloke, he had a spot of trouble from flashing and doing robberies when he were young. Scared they'd try to put it on him because of that record, so he talked *me* into it. I didn't think they'd bother with me. I didn't even live in the bleedin' village when that first girl got killed. But now they sent *me* a letter too! I'm in trouble because of *him!*"

Ian told Colin he was sorry to hear about his troubles, but didn't know what to advise. Ian later said, "If he'd offered me two hundred quid or something I'd have thought there was something wrong, but he didn't say that. He just said I could do it for him. That I got nothing to worry about."

Then toward the end of January, Colin took Ian Kelly aside again. This time he was more desperate.

"The twenty-seventh!" Colin said. "I got to give the blood on the twenty-seventh! Why *wouldn't* you do it? I did it for the other bloke and I didn't know him near as well as you know me! Now I might get nicked if they find out I gave blood twice! Look, I got kids. You don't have no kids. You can keep me from getting in trouble!"

"He had them kind of eyes," Ian later said, "eyes that look like they could kill you."

Ian said to Colin, "Okay, get off me back, I'll do it."

"We got to go to a photo booth," Colin told him. "We got to do it right."

That afternoon Colin Pitchfork drove in his Fiat with Ian Kelly to the photo booth at the Leicester railway station in London Road. They took a strip of passport-sized photos.

But on the day of bloodletting, Ian Kelly became ill. His temperature shot up to 103 degrees, yet he somehow made it through the workday with the help of medication he'd gotten from the chemist's shop.

"It's like I'm in a bloody sauna!" he told Colin that afternoon. "I got me a bleedin chest infection, mate. I'm sick!"

"Just hold on long enough to do the test tonight," Colin told him. "*Then* you can go to bed and take your pills."

Later that afternoon when Carole was at work, Ian
Kelly reclined in Colin Pitchfork's living room trying to
watch television while Colin sat at the table working on
his passport with a razor blade. He cut around his
photo and slid it out from under the embossed plastic
lamination. He cut Ian's photo slightly larger and slid it
back inside the stamped plastic shield. He pressed the
edges with a bit of liquid sealant and it looked surpris-
ingly good.

"Bingo! It's done!" he said to Ian.

Then Ian was schooled on answers to questions about
the names of Colin's kids and when they'd been born.
But Ian was coughing too much, was too fever-wracked
to get any of it into his head. He went home to rest
until it was time.

The Pitchforks had a friend and neighbor named
Mandy, an exuberant young woman who'd always found
Colin to be an easygoing fellow and very fond of his
children. At a later time she said of him: "On numerous
occasions he made minor passes at me and invited me
to bed, and pushed his body and private parts against
mine when he passed by. I always treated these incidents lightly, and he often called round my house as a
friend."

On the 27th of January she agreed to baby-sit while
Colin went off to be bloodied and Carole went to school.

Early that evening Colin arrived at Ian Kelly's home
in Leicester, went up to the bedroom and privately
talked his ailing friend into getting up from his sickbed.
Ian's temperature was approaching 104 degrees, but, as
Carole always said of Colin, he *was* persuasive. Five
minutes later they went out the door together. It was
the first and only time Colin would ever visit the Kelly
house.

When they got to Danesmill School on Mill Lane in Enderby that Tuesday night, neither man noticed how cold it was—Ian because he was weak, light-headed, burning with fever; Colin Pitchfork because he was more worried than he'd ever been in his life.

*After he went inside I sat in the car for five minutes and then I thought, This is bloody daft! I moved the car round the corner out of the way and left it. I walked round the block up to the main road, back down, then back up to the playing fields so I could see across. So if suddenly a police van or car shot out there toward the chip shop I'd know something was wrong!*

He waited in the darkness. The school was in the same street where Dawn Ashworth had lived. Just down the road was a lane leading to a footpath, to the place where she'd died. And inside the school, policemen were waiting for his *blood*.

"There were rows of coppers in there," Ian later recalled. "All in civilian clothes. I were shaking like a leaf from the fever. I sat down and waited."

Ian hardly realized it when the name was called. "Colin Pitchfork!"

"That's me," Ian Kelly answered.

Ian sat at the table facing the detective and signed Colin's name and filled out a form. A detective sergeant examined the passport and driving license. There were quite a few young men there because the blooding was only in its first month. It was a valid passport so no Polaroid was necessary. The consent form was signed by Ian and he was escorted to a physician who took the blood and saliva samples, to which the detective attached labels.

Colin Pitchfork had thus officially complied voluntarily by giving samples of blood and saliva, proving his identity with British passport number P413736B, along with a valid driving license.

When Ian Kelly emerged from the school, Colin Pitchfork watched and waited, but left Ian shivering in the road until he was sure it was safe. Then Colin emerged from the shadows and called out. When they got to the car, Ian Kelly gave Colin Pitchfork the passport and wallet.

"Everything all right?" Colin asked.

"Yeah," Ian answered. "Doddle, dead easy."

Colin drove Ian Kelly to his house in Leicester, dropped him and said, "Cheers, mate. And mum's the word."

Before Carole got home from school that evening, Colin had scratched a mark on his inner forearm with a compass point, and stuck some adhesive plaster over the wound.

When she saw it, Carole said, "I thought they'd just prick your thumb."

"No, they used a needle and jabbed it in me arm!" he said. "And I didn't like the attitude of them coppers. Treated me like a criminal, they did."

Then he made a big fuss, as he always did when removing an adhesive plaster. He couldn't bear the pain. He clipped around it with a scissors and pulled it off as if removing sutures. He showed her the wound.

"Look at that," he said. "And they made me chew on a piece of cloth. Me bloody arm is killing me."

"You baby!" she said. "A teeny pinprick."

"They took a photo cause I only had a driving license," he told his wife. "They may bring it round here to show people to prove it's me."

When the hand-delivered letter arrived at the little house in Haybarn Close saying that Colin Pitchfork's test was negative, essentially eliminating him from the murder inquiry, Carole Pitchfork showed the greater relief. In fact, her relief was greater than she dared admit to anyone, even herself.

There wasn't much time for Carole to savor the small victory of forcing Colin to take that blood test. He began wallowing in depression again.

"What's wrong with you?" Carole finally asked him.

It was a few days after the test. He sat staring and sighing. Her way, as she put it, "was just to fly off the handle and cause a row when he was like that." So she did: "For chrissake, just tell me what's wrong, whatever it is!"

He looked up at her with tears welling and admitted his affair with Brown Eyes. "She had a baby," he said, "and it were stillborn! *My* baby daughter!"

Carole was never sure what she said next, if anything. She slammed out the door and went to Mandy's house, not returning until evening. Locked out of the bedroom, Colin slept on the settee that night.

The next day she had to talk to him about it for the sake of her sanity.

"I can't *believe* it! This woman is abrasive. She's domineering. She's not your sort at all. She's thirty years old if she's a day. By your standards she's a bleedin hag!"

"Didn't you ever suspect?" he asked curiously.

"It crossed my mind. But no, I couldn't ever believe that. Not with *her*."

"Look," he said, trying to be reasonable, "it's been going on for over a year and you didn't know, did you? It's not doing you any harm then, is it?"

"One question, you bastard. Did she ever hear of the pill?"

"Forgot to take it a few times," he said.

"Yeah, for a year. Did she forget *every* time?"

"Maybe she wanted to trap me," he said.

Later, Carole Pitchfork thought about it. About him, about herself, about the way he viewed things.

"I actually believe he had this *rosy* picture," she told a friend, "of us taking that woman into court for a grand fight to get custody of his baby. He probably had a fantasy of bringing baby home where we'd all live happily, with me as the mummy."

"Get the hell out of my life," she told him, and he did.

When Colin was gone the phone calls started. Brown Eyes began ringing Carole and writing notes. It got so bad that whenever the phone rang, Carole hated to pick it up.

"Are you sure you know where Colin is?" the telephone voice would say. "He might be here with me. Anytime I want him, he *will* be here with me!"

While Brown Eyes recuperated at her parents' home, Colin went to visit her daily. He'd stay for half an hour and talk sadly about the baby. He offered to pay for the funeral and they buried the infant at Wigston Cemetery. Colin stood at the grave and wept.

Three months later, Brown Eyes' mother arrived at her house unexpectedly and caught her daughter and Colin Pitchfork in an act of sex down on the living room carpet. The outraged woman took her grandchild home to her own house and made her daughter promise never to see Colin again.

The interlude was over. Colin Pitchfork had seemed to relish living a soap opera, but the star-crossed bakers were even cautioned by their foreman at work. Brown Eyes never saw Colin outside of the job again.

She referred to him as a very gentle person.

*     *     *

During their estrangement Colin visited Carole and the children frequently, and even babysat when she was at work or school. The children missed their father a lot, and he wanted to come home.

"The idea of you seeing other men actually eats me up!" he told Carole. "Now I know what's what. I've learned a lesson."

"I weighed it," she later remembered. "The effect his absence had on the kids. And as usual . . ."

When Carole let him return home in March they still slept apart. And she began monitoring the mileage on his car and motorbike, writing down the numbers on the odometers. When there was too much mileage, he'd always have some plausible excuse.

It didn't seem possible to her that he could be going out "on a wander," looking for girls to flash. He *couldn't* be regressing to that, and yet . . .

Carole Pitchfork never knew how easy it was to disconnect an odometer temporarily. Nor did she understand that there was another world out there belonging only to *him*, a world of heightened reality. While he lived out a mere shadow life with her in Haybarn Close, during those phantom days.

# 23
# Bloodprint

I am in blood
Stepp'd in so far, that, should I wade no more,
Returning were as tedious as go o'er.
—*Macbeth*, Act III, Scene 4


They had a dozen GP's taking turns at the blooding, some of whom also practiced as police surgeons and dealt with sick prisoners, Breathalyzers and police medical exams. The inquiry got them on the cheap, at £72 for a two-hour blooding, and a good doctor could bloody forty men.

The murder squad had fine-tuned cajolery by telephone. You could walk into the incident room at any hour of the day and hear them at it:

"Cornwall isn't *that* far. What kind of car do you drive?"

Or "Don't you want to visit the girl you left behind?"

Or "Wouldn't your mum just love to see you?"

Derek Pearce called the out-of-town trips "speed runs." "We did them at various doctors' surgeries around the country and knocked off four to five hundred miles a day sometimes, just to get a few blood samples. Since we weren't given funds to stop overnight, we had to put our foot down to make it back home at a decent hour."

Since they got no overtime pay, even for working on rest days, and no meal allowance, the people of Leicestershire got an inexpensive investigation, given the massive scope of it. As Pearce put it: "Our goal in life wasn't to make money. It was to detect murder. *These* murders."

On many evenings when Pearce had to stay late because of some tedious management task, he'd find one or more of his detectives dropping into the incident room after supper.

"To sort out a thing or two," they'd tell Pearce.

Some of them played hard. One of the policewomen who worked on the inquiry had what she apparently thought was a *private* understanding with one of the detectives. She'd leave for twenty minutes, usually in the midst of a boring shift when everyone was busy with paperwork. Each time she disappeared, a certain detective would receive a telephone call at his desk, a *lengthy* telephone call.

Sitting there among fifteen other detectives, he never had too much to say from his end, but his eyes would roll and flutter, and he'd get a cunning little grin on his face. And he wasn't fooling anyone when he'd say things into the phone like "Yes, I think we can accommodate your needs quite nicely. Always glad to help. Cheers, mate!"

After a while, the others decided to let it be known that it couldn't be pulled off right under the noses of veteran police detectives. One night when they all went to the pub for a little do, the policewoman was presented with a gift from the squad: a pair of empty soup tins with the tops removed, joined at the bottom by a fifty-foot string. One of the cans bore the number of the male detective's telephone, and the other was labeled with the number of the "secret" telephone terminal.

"To save steps and shoe leather," the gift givers said.

\*        \*        \*

There were a sizable number of objectors who could be eliminated on paper after their alibis were checked, but many of those who refused did so not because of personal liberties but because they were *terrified* of needles. The saliva test was all right as a backup, but it didn't provide what the first-phase blood grouping required, and wasn't always good enough for a proper DNA analysis.

One of those who rang in refused to be stuck with a needle but was otherwise cooperative. "Look," he said, "I want to help! I'll come in, but I'll poonch meself in the nose!"

Another said, "I won't have the needle, but I'll let the doctor cut me with a knife. I don't mind the knife. I'll even cut meself with a knife!"

Without consulting Freudian textbooks the cops just accepted that there were people with a needle phobia, so Pearce agreed to a cutting for one of them.

When the donor showed up, Pearce was astonished: "He was a big strappin bloke! Sixteen stone, and stood six foot up!"

The strapping bloke with the belonephobia was taken to a doctor who said, "We'll cut you and squeeze a few drops onto the card. We'll screen it for a certain blood factor and if it comes back negative, fair enough. But if it comes back positive, you'll come back in for a proper test, okay?"

"Anything! Anything," the donor said. "But I'm *not* having a needle!"

The doctor said, "Okay, turn your head away and I'll cut your finger."

The mountainous donor turned away and gritted his teeth while the doctor cut his finger and took the blood. The situation was resolved. In that he had no scalpel with him, the doctor made the cut with a needle.

\*   \*   \*

Some of them adamantly refused and wouldn't budge. In fact, an acquaintance of the kitchen porter, who openly resented the way he had been imprisoned by police, refused to come in at all. One of the sergeants went to Derek Pearce and said, "He won't give one."

"What do you mean, he won't give one?"

"He says it's an infringement on his personal liberties."

"Come on," Pearce said, taking the detective and driving to the young man's house.

The objector allowed the cops to enter. As Pearce described him, the young fellow was "dirty and ugly and grimy. The kind of bloke you'd like to take out in the field and shoot."

Pearce didn't shoot him. He talked. His subordinates said that Pearce was "the kind who could talk the knickers off a nun." Even so, it took an hour before the young man promised Pearce he'd come in for blooding the next night. Several detectives made bets that he wouldn't show, but he did.

When he arrived he was just as dirty and ugly and grimy as he'd been the day before. He was also surly, with a total vocabulary of about seventy-five words. He grudgingly answered the questions, filled out the form, walked over to the doctor and disdainfully watched the blood being sucked out of his arm. Then he keeled over on the floor. Out cold.

They ordered an ambulance to take him in for observation, and a detective was detailed to remain at his bedside until he came around. In a piece of profound understatement, Pearce said, "No other police force *ever* had to do anything like this before."

They had their fair share of people with AIDS and hepatitis. When they knew it in advance, they'd take the saliva tests *very* carefully. The gauze for the saliva was attached to a folded card, and was designed to drop out so that the subject could catch it in his mouth. They couldn't get people to do it right. Some sucked on the gauze. Some sucked on the card. Some chewed the gauze. Some chewed the card to pieces. Some nearly *swallowed* the gauze and ended up gagging. It was a disgusting business, they all agreed.

Angry young women would come in with boyfriends and say, "I want him *cleared*!"

One girlfriend said, "Can you test him for syphilis while you're at it?"

The cops who delivered the letters to the houses used to amuse themselves by telling the donors, "The good news is you haven't done the murders. The bad news is you've got AIDS."

They bloodied quite a few young policemen who lived in the villages, and the new headquarters building was full of officers who worked the motorway around the three villages. Derek Pearce, who'd had a few problems in his day with traffic officers, couldn't wait. "I want to bloody the traffic cops," he always said. "With a very blunt needle."

Halfway through the massive testing Pearce *was* involved in a personal blooding. He went along on one of the out-of-town treks to pick up a young man who'd agreed to be bloodied at the surgery of a local physician. They collected the nervous lad and drove him to the doctor.

The physician wasn't exactly the doctor-priest gowned for surgery and tended by starched minions at the altar of Hippocrates. This one was the kind who might treat a sick cow if the vet was drunk. Pearce handed the rumpled country doctor a syringe, swab and plaster from the test kit. The doctor had never seen a self-sealing syringe. He couldn't manage to attach the needle. He couldn't locate a vein.

"He had three or four goes with the needle," Pearce recalled, "but he couldn't find a thing."

The victim didn't say much. He just sweated, and muttered, "Blimey."

The doctor dropped a needle on the floor trying to attach it. Then another. He couldn't figure out how to unscrew the plunger in case he ever *did* get any blood. He made a few more stabs. He started taking divots.

In the end, Derek Pearce became the physician. The two of them slid the needle into what looked like a vein and Pearce held the syringe while the doctor pumped and the donor said, "Blimey."

"We'll send you thirty quid for this," Pearce told the doctor after putting a vial of blood safely in his pocket.

The donor looked like he'd been dueling with Jack the Ripper. "Thirty quid for *him*?" he said. "I should get sixty!"

"You get our thanks," Pearce assured him. "Isn't that enough?"

"Blimey!" said the donor.

Pearce later described the donor as "now needing a prosthesis."

Everyone came to give blood. A former patient from Carlton Hayes Hospital came because his name had turned up on the computer list. The reason it turned up was that he'd gone to the hospital at the time of the Dawn Ashworth murder to borrow crutches! They took his blood anyway.

A transvestite came wearing a red lamé dress. They took *her* blood.

They'd gotten television coverage in Australia, Brazil, the United States, Sweden, France, Holland and had been on the air live in Italy. One night a West German camera crew came to watch the blooding, and interviewed two detectives on videotape.

"Does the process in any way hurt or cause anxiety?" was the on-camera question.

"Absolutely not," the detective said to the camera.

"Do the young men mind giving blood?"

"Absolutely not," the detective answered.

The next man to step before the camera took one look at his blood being siphoned, and keeled over on top of the detective who kept saying "Absolutely not." He was carried out by two bobbies while the camera rolled and the blood flowed, camera or not.

The laboratory at Huntingdon was freezing just about all of the blood by then, and not sending it to Aldermaston for the DNA test. The lab asked the murder squad to stop sending more unless it was "high priority." The laboratory spokesman informed them that the technicians had done a year's work in a few months and were stretched to the limit. They were drowning in blood. There were vials on every shelf. The freezers were full of it. There was more young British blood flowing in Leicestershire than had been spilled at the Somme.

But nothing stopped them. The murder squad sought blood tirelessly. It was their best and only hope. The mere *act* of the blooding might cause the killer to bolt and run.

They bloodied them all: transvestites, policemen, limpers with borrowed crutches, *anyone* who fell within the age group. But probably no blooding was as strange as the sample they took from the "Very Reluctant Donor," who said that he would not, absolutely, positively, unequivocally, under any circumstances, surrender a drop of his blood whether they took it with a needle, or a scalpel, or a machete.

He lived in a filthy flat and nothing would move him from his hovel. It was a dilemma worthy of Derek Pearce and maybe an envoy of the archbishop of Canterbury.

Pearce had an idea, rang the Very Reluctant Donor, and said, "How about giving a *semen* sample?"

"I don't mind *that!*" the man replied enthusiastically.

The next morning they took him from his filthy flat to a moderately clean doctor's room at Wigston Police Station.

"They stripped him off, tossed his greasy sou'wester in a corner and put him in a track suit," Pearce remembered, "so he couldn't take anything in with him. Wigston was packed like a railway station that day, and he took *ever* so long in that room alone."

Finally, the Very Reluctant Donor knocked and the cops opened the door. "It's no good at all!" he cried. "I'm trying me hardest and I can't do it!"

The cops huddled again and a few silly ideas were tossed around, but DC John Reid had a thought: "Would a *book* help?"

"Oh, yes! Ever so!" said the Very Reluctant Donor. "With pictures?"

So a detective was sent to the custody room to see if they could locate a good one, with pictures. Somehow the message got garbled and a magazine was brought to them called *Locomotives and Railway Time Tables*. And it wasn't even a current edition.

"We'll try again tomorrow," Reid said in frustration.

"No, no! Give me another go!" the Very Reluctant Donor cried, and went back inside the room for a *very* long time.

He eventually knocked on the door, emerged, and

proudly extended his palm. Resting on his filthy fingers was a specimen card bearing a filthy little glop of something or other that made Reid want to *retch*. It had taken the Very Reluctant Donor exactly thirty-two minutes to stand and deliver. He seemed humiliated by the fact that so much work had brought forth such little product.

He explained it by saying, "See, before I left home I *did* one for you, ever so much better! But I forgot to bring it *with* me!"

Some of those who refused to be bloodied did so because they had what they believed were definite alibis. One of those offered to produce several friends he'd been with at the time of Dawn Ashworth's murder.

The newspaper picked up on the issue and Chief Supt. David Baker was forced to issue a statement saying, "There is no reason to suspect him any more than anyone else."

A Midlands television panel presented a debate and examined a larger question: "Does mass screening for murder pose as big a threat to civil liberties as it does to killers?"

An attorney for the National Council for Civil Liberties, a guest on the show, thought it did. He said, "The police letter claims that the test is voluntary, but it implies that if you don't telephone for an appointment there'll be a knock on your door."

The television reporters went out onto the streets of the villages where the snow was piled in one-foot drifts, and interviewed young men. Most thought that the test was worthwhile, with certain reservations. The last young man interviewed spelled out the reservation.

He said, "*Their* person ain't gonna go in, is he? The one that's done it?"

Another worry of the civil liberties lawyer concerned what the police would do with the DNA information. Government proponents of the mass screening had pointed out that the Police and Criminal Evidence Act required that samples be destroyed if proved to be negative.

"They might destroy samples of blood but not destroy the information *taken* from the samples," the civil liberties lawyer said.

When Supt. Tony Painter was interviewed he reiterated that there was no undue pressure on young men, that the tests were completely voluntary. He promised to "stay within the guidelines."

The civil liberties attorney countered that the Home Office *had* no guidelines for this kind of mass screening, that genetic fingerprinting was so new nobody had considered its larger implications.

Then he talked about a "national bank of DNA" and said, "If we're going to do things like this we could end up fingerprinting everyone at birth! It smacks of Big Brother. We need a privacy act to safeguard us."

Some of the young men in the studio audience agreed, and resented answering the police questions that accompanied the test. One of them disliked being asked to report in the first place, since he hadn't even lived in the village during the first murder.

Thus, he presented the *same* argument that workers in Hampshires Bakery had heard from Colin Pitchfork.

One of the regular doctors who did the blooding was an old chap who couldn't see very well and who liked his drink, and was a bit shaky from liking it so well. The cops noticed early on that if a doctor missed a vein four or five times someone might get queasy. More than five stabs, and they could get ready to call the ambulance.

If a polite young man came in and said, "I have this old army photo you can borrow," or "It's perfectly okay to take a photo," they'd send him to one of the younger doctors.

If he came in and said, "I don't see why I should have to put up with this!" he was directed to the old doctor.

John Dayman, the Duke Wayne impressionist, was sent out one evening to pick up an Asian donor who'd agreed to come in. As the cops well knew, the Asians wouldn't go anywhere unless the whole family went, so Dayman used the "Asian ghetto car" for this one—"the biggest car in the world."

Sure enough, when he arrived at the donor's house he was told, "Taking wife."

Dayman said, "Yeah, yeah, I was expecting that. There's plenty room."

"Kids got to come if wife come," the donor said.

"Okay, mate, bring the kiddies," said Dayman.

"Brother-in-law come."

"Now wait just a . . ."

"Mother come," the donor said. "And mother's brother come."

"How about your bleedin grandfather?"

"No grandfather. Auntie. Auntie come," the donor said.

Pretty soon they were all in the car piled on each other's laps. "The bloody car looked like a motorboat with the lights pointing in the air!" Dayman said later.

While en route the donor said, "Police car no good. No automatic. My car got automatic."

"I'm ever so sorry you don't like the car," said the cop.

"Radio no good," the donor said. "Get music on

radio. No music? I got Datsun. Good car. Automatic. Good radio. Got fag?"

"Yeah, I got fag." Dayman gave the donor a cigarette.

"Brother-in-law want fag," the donor said. "Wife want fag."

"Everybody want bleedin fag!" Dayman said. "Here, take the lot!"

Before arriving, Dayman noticed that the kid next to him had wiped his nose on the seat. When he got the family to the blooding, with everyone smoking his fags, demanding Polaroids of the kids, complaining about the stale tea, Dayman scraped the snot off his coat, went straight to an interviewing detective, pointed at the half-blind doctor, then to his donor, and said, *"Hurt him!"*

Another time, John Dayman, who'd had prior experience wrestling maniacs, was detailed to pick up three former patients from Carlton Hayes Hospital: two in Coalville, one in Ibstock. While driving the three men toward the blooding center late in the afternoon, he became uncomfortably aware that the passengers' conversation was growing more stressful with each mile, as they got nearer to being bloodied and nearer to their former place of confinement.

He thought he'd try to take their minds off it. The detective who'd made the last run had left a stack of tapes in the car, so Dayman said, "Anybody care to hear some music?"

Without waiting for an answer he grabbed a tape at random and shoved it into the player. It was music from Monty Python's "The Idiot Song." Thirty seconds after he punched the button a voice sang: "Fee fi fo fum! I smell the blood of an a-sy-luuum!"

The passengers grew silent. Dayman's hand crept toward the off switch.

Since most donors didn't have good identity cards and had to be photographed, the police needed a train-load of Polaroid film for photo identifications. Often they took Polaroids of the kids to keep them amused while the old man was being bloodied. They posted one in the incident room of Derek Pearce baby-talking an infant who wasn't all that amused by the bearded nanny.

Sometimes a donor objected to a Polaroid even more than to the needle. One evening a prison officer's son came in with his mother. She was irate that they insisted on the photo and didn't accept her word that the lad was her son.

A detective later carried the photo to a neighbor in Littlethorpe who did indeed verify the young man's identity. The detective thanked the neighbor for being of assistance.

"Glad to help," Colin Pitchfork told the policeman. "Cheers, mate!"

# 24
# Anniversary

The psychopath is a hedonist, a pleasure seeker. Self-pleasures and satisfactions are very important to him.

. . . if a fancy or whim passes through his mind, it becomes quickly converted to action. Possible negative consequences of his acts do not concern him. Rather, he has a need for stimulation and acts recklessly, thoughtlessly taking risks, sometimes harming others, and not thinking about future consequences.

—RIMM and SOMERVILL

By the bloody month of May, the murder squad had bloodied 3,653 men and boys, yet only 2,000 had been eliminated, due to the workload under which laboratory technicians labored. The police had by then received an incredible 98 percent response for the voluntary testing. Yet it seemed that one donor kept leading to another, and they always learned about new ones.

They ultimately decided to bloody everybody, regardless of alibis. Police estimated that they had about 1,000 *more* men to contact, and they were down to twenty-four officers.

Aside from police work Derek Pearce had one passion: cooking. Mick Thomas and Phil Beeken—and a few others who genuinely appreciated his efforts—used to drop by for his stuffed trout, or his chicken in wine sauce stuffed with leeks and Stilton, or prawns in brandy and cream sauce with lots of garlic. He went to great

293

lengths with his salads, putting little teeth in the radishes, and feathering the tops of spring onions so they'd hang over just so. The presentation mattered as much as the taste to this perfectionist.

Pearce's kitchen, which he'd built himself, was spacious, with lots of cupboards full of spices. He was keen on Indian dishes, kept twenty kinds of curry ingredients, and was always eager to marinate chicken in yogurt with *tandoori* sauces. Pearce prided himself on having several delicious and sightly meals that he could "whack up quickly" if a friend popped in.

Sometimes after a great meal, and wine, and drinks after dinner, he could even relax enough to lower his defenses. On those fleeting occasions when Derek Pearce alluded to his ex-wife he'd unwittingly reveal himself. In that he was so often described as "abrasive" it was surprising on those relaxed occasions to hear him say shyly, "I know I'm not good-looking, and I'm *very* hard for a woman to live with."

It was rather touching because he obviously meant it, perhaps explaining the full beard. Yet his self-assessment wasn't accurate. Without the theatrical beard—more important, without the look of driven intensity—Pearce could be called attractive. His features were regular and strong. He had good, dense dark hair and expressive eyes. His nose was well shaped, with a slight character-giving bend. He had a boyish smile, was a glib, enthusiastic talker and was very popular in the secret pubs he shared with no other cops.

Though he couldn't stand to be at rest, he claimed never to suffer from stress.

"Give me a problem," he said, "and you've given me *life*! I have to burn off part of me so I can put my head down on a pillow at night. Four games of squash don't help. I need *problems* to solve."

Pearce lived with acrophobia so severe that he felt unreasoning terror on stepladders and even open-air buses. Sometimes he'd talk about it. During those telling glimpses, one could conclude that the fuel powering Derek Pearce was pumped from a well of insecurity, causing the behavior that made them say, "You either like him or you don't."

Pearce certainly had never endeared himself to lady friends when he said things like "No matter how much I care for a woman, I'd rather go to work."

He was so defensive he kept his emotions combat-ready, occasionally making preemptive and unprovoked sorties on the world around him. Yet the man who was so difficult to live with *hated* to be alone.

Those colleagues who liked him seemed instinctively to understand that his insecurity was at the heart of his conduct. Those who didn't like him . . . well, there were many of those, some in high places on the police force.

A fellow detective on the murder squad said of Pearce, "When Derek's not fully occupied he gets in a bit of bother. He needs a lot of outside pressure or he gets bored and then he gets himself into dodgy places." He was to get himself into a *decidedly* dodgy place.

During the Lynda Mann inquiry a female police trainee caught his eye, when he watched her walk. She was then only seventeen, but two years later she joined the force and was assigned to Braunstone Police Station, under the jurisdiction of South Division CID. This put her technically under the jurisdiction of Derek Pearce. The fifteen-year difference in their ages, and the difference in their police ranks, made a romantic relationship a bit dicey.

On an evening in May 1987, Derek Pearce and Insp. Mick Thomas were called to the home of Supt. Tony

Painter to discuss staffing levels. At the end of that
meeting Pearce made a decision to visit the young
policewoman at her Braunstone flat. It would rank among
the worst decisions of his life.

One of Pearce's thirty-five-year-old lady friends tried
to explain it to him. She said, "You can't mess with a
twenty-one-year-old's feelings the way you do with
*mine*."

Pearce looked enough like a repertory company
Petruchio that maybe he decided to play the role in
earnest on that May evening. The local newspaper carried the story:

### POLICE ENQUIRY INTO OFFICERS' ROW

A disciplinary enquiry has been launched by
Leicestershire police after an incident in which a
female officer was hurt during a row with a detective
inspector outside her home.

Distressed neighbours called the police to the
incident in Braunstone. When they arrived they
found Det. Insp. Derek Pearce and WPC Alison
McDonnell arguing.

WPC McDonnell, who has lodged a complaint
against Det. Insp. Pearce, is understood to have
suffered minor injuries to her face, and the front
door of her home had been damaged. A complaint
has also been lodged by the owner of the house,
WPC Elizabeth Pell.

A spokesman for Leicestershire police confirmed
today an incident took place leading to disciplinary
proceedings. Neither officer was being suspended
while enquiries were made.

He said the incident happened more than three
weeks ago. Mr. Pearce, who is aged 36, and and
divorced, is one of the leading officers in the

investigations into the murders of schoolgirls Dawn
Ashworth and Lynda Mann. Dawn was found dead
in Enderby last August, not far from the site of
Lynda's murder in November 1983.

WPC McDonnell, aged 21, is based at Braunstone Police Station and was featured in the
Leicester Mercury two years ago when—just seven
weeks into her police career—she helped deliver
a baby.

Neither officer could be contacted for comment
today.

Referring to the quirky charm of Derek Pearce, one
of his men said, "He was the kind who could grab a
handful of daffodils and make antagonistic people respond to him."

But this time there were no daffodils. Criminal charges
were initiated. Pearce wouldn't talk about the case at
all, except to say he was innocent.

He continued to work on the murder inquiry, though
in a rather more subdued fashion, while the internal
investigation into his own case was quietly under way.
Pearce was a valuable commodity, especially in this
critical stage of Dawn Ashworth II.

In early June, a seventeen-year-old girl from Oadby
spent the evening with friends in Wigston Centre. She
got into an argument with her boyfriend, left him, and
decided to walk home. It was just after midnight when
a blue car pulled up beside her.

The driver said, "Where you going, m'duck?"

She looked inside and asked, "Are you going to
Oadby?"

*I'd only had three hours' sleep the night before, but I
had to go out that night. Carole was gone with the kids*

*on a Saturday night camp. When I got on a high like this I had to drive around. Sleep and fatigue just didn't matter. You become superhuman! So at midnight I went out for a wander. I drove through the center of Wigston and saw a young girl saying good night to another girl and a bloke. She walked off, and I drove round the block and come up to the roundabout and out comes her thumb! I thought, "Fuckin hell, Colin! This is your lucky night, ain't it?"*

"I'm going up the A-Six," he said. "Any help?"

"Oh, superb!" she said, and hopped in.

"What's your name, m'duck?" he asked.

"Liz," she answered.

*I knew she was no older than eighteen, blond and full of bounce. Lives with Mum and Dad. Been out with friends for the night. Just the type!*

She secured her seat belt and they rode for a few minutes, but Liz started getting very nervous. He was expressionless and didn't speak anymore. She hadn't liked his menacing grin.

"Where do *you* live?" she asked.

"What?"

"You said Oadby was on the way to where you were going."

"No," he said. "No."

They were driving to the center of Oadby then, to the junction of the main A6 road.

"That's the turning there!" she said.

But he silently drove past it.

"There's a turning here!" she said. "Go back!"

Still he didn't reply.

"I want to get out!" she cried.

He slowed for a second, but only to change gears, then he sped off, heading for the countryside.

They passed a pub and she screamed, "There! Turn in!"

But he was driving down a dark country lane.

Suddenly she grabbed the steering wheel and he had to hang on and mash the brakes!

"We'll crash!" he yelled as the car skidded to a stop.

"I thought you wanted it!" he said, while she sobbed hysterically. "I thought this is what you wanted!"

"All I want is to go home!" she wailed.

He started the car warily and turned around in a field entrance. His whole demeanor changed. He said quietly, "I've had a drink or two, you see."

"I can drive if you've been drinking!" she sobbed.

He put his hand on her knee and said, "I've not hurt you yet."

When they got near the A6 he pulled over and quickly opened her door. But he held on to the handle. "Give me a kiss then," he said.

The girl threw herself against the door and leaped out, crying and running.

He shouted into the night, "I bet you'll never accept a lift again!"

When she talked to police about the incident she said, "He had unusual staring eyes. Like *dead* eyes."

The murder squad was cut back to sixteen officers in all, and both inspectors—Derek Pearce in his hot-blooded confrontational style, and Mick Thomas, cooler, more detached and businesslike—battled to keep the top brass from shutting them down. Though it was decidedly unpolitical, Pearce let it be known that if the chief constable's office tried to close up the incident room and disband the squad, he was going to the press.

Yet they had tested four thousand men, and undeniably, the budget was stretched. A newspaper headline said: NO LEADS TO KILLER OF LYNDA. It was followed by a huge story on the 31st of July:

**DAWN'S KILLER IS STILL AT LARGE**

The family of Lynda Mann was contacted in Lincolnshire and Kath Eastwood said, "It's a process of elimination at the moment. I don't think they can do any more. I'll never give up hope and I'm sure they will find him in the end."

Supt. Tony Painter issued a statement saying, "Dawn was murdered on July 31st and we mean to use the anniversary to give the enquiry another boost. We know there's a risk that this evil man will strike again, and we know that there's information in the community that could lead us to him."

It was the same old story. The police were asking for help and getting nothing of value.

For nearly a year Dawn Ashworth's grave had not been marked by a headstone, and the Ashworths were getting reports that Dawn's friends kept placing flowers on various wrong graves. They decided to get a headstone, and secured a loan to do it. They chose one made of black marble, with a carved gilded path winding toward a sunrise. The inscription said:

Treasured memories of our dear daughter
**DAWN AMANDA ASHWORTH**
Born 23rd June 1971
Tragically taken 31st July 1986
What we keep in memory is ours unchanged
forever

"It was all we could do for her sixteenth birthday," Barbara Ashworth said.

The Ashworths were asked to pose as a group for another news portrait that summer. It was nothing like the one taken in front of the bay window when Dawn was alive, when all of them had linked arms and beamed at the camera. In this photo the three surviving family members looked soberly at the photographer. The extraordinary thing about that photo was that Sultan, their English setter, perhaps reflected his family's emotional vibrations. Their story was written in the dog's face.

Sitting at Barbara's knee, Sultan posed patiently, like the others, but with slightly averted eyes. Eyes that looked utterly grief-stricken.

Prior to the anniversary of Dawn Ashworth's death the police put posters in the shops and on all the notice boards in Narborough, Littlethorpe and Enderby. But Carole Pitchfork didn't hear much talk about the murder anymore. Nobody speculated about the picture of the punk with the spiky hair. Carole's friends and neighbors seldom bothered to cast back their minds to remember someone who was "badly marked" from a death struggle with Dawn Ashworth. Villagers stopped speculating whether or not a wife or mother or father had taken the killer's bloody T-shirt and buried it in a garden.

The only thing that Colin Pitchfork had to say to his wife on the subject of the reinvigorated murder hunt was "You'd think they'd have left the posters up all year. You'd think they'd make more of an effort."

When the anniversary of the murder was approaching, the murder squad wanted to do a covert operation, what they called a "discreet observation" of the gravesite in the churchyard of St. John Baptist.

Derek Pearce visited the vicar of Enderby one afternoon, accompanied by DC Phil Beeken, a tall, handsome young fellow who was a friend to Pearce both on and off the job.

Pearce needed to find a good observation point from which to watch those who might pay a visit to the grave on that occasion, but there wasn't any. Putting someone as large as Phil Beeken out there in the middle of the churchyard wouldn't do. Beeky would be about as inconspicuous as a solitary tooth.

Then Pearce noticed that the graveyard was in a terrible state, all overgrown with grass and weeds. So he made the vicar of Enderby an offer he couldn't refuse.

He asked the vicar, "How would it be if one of our lads tidied up the graveyard and cut the grass for a week or so? No charge!"

"I'll do it for you, boss," Beeky said to Pearce. "If it's okay with the vicar."

The vicar was enthusiastic, but before allowing Beeky to go to work, he insisted on demonstrating the use of a grass-cutting "strimmer." While he was demonstrating it to the cop, the vicar strimmed Beeky's trousers and the leg inside. He apologized profusely to the young detective who told him not to worry, it wasn't bleeding *all* that much.

Phil Beeken immediately became part of CID trivia: Who was the only detective ever to be wounded on duty by the vicar of Enderby?

Beeky had volunteered for graveyard duty on a bright sunny day. But when he went to work as a cemetery gardener it rained all week. He had a hand-held radio, but not one that worked, so he usually had to brave the rain to monitor the movements of mourners. Frequently they offered him money for tidying up the graves of loved ones. He didn't accept the money, but made many friends while getting soaked to the bone.

They'd put a video camera on Dawn Ashworth's grave, just as they'd done on Lynda Mann's grave during the various anniversaries since she'd been murdered. It was a time-lapse video, and though they studied the tape they never saw anyone who might be *him*.

Yet one of the visitors to Dawn Ashworth's grave did stir a bit of notice. He was a salesman up from the Thames Valley who had half an hour to kill, and had decided to take a stroll through the old churchyard. By pure chance he stood at Dawn's grave, and was swooped on by police observers.

The salesman was interviewed and released, but the police in his hometown were contacted and asked to verify his reputation. As bad luck would have it, his local constabulary was also investigating the murder of a schoolgirl, so the salesman got grillings on that murder *and* the Dawn Ashworth killing before the police were satisfied he was innocent.

He vowed that he'd never enter another graveyard, alive.

The local cops told him he was lucky those Leicestershire ghouls hadn't taken his blood.

Nine days after Dawn Ashworth's murder they'd questioned a patient at Carlton Hayes Hospital, a man twenty-five years older than any other supect on their list. He had previous convictions for indecencies and was one of those few who'd been blood-tested back in 1983 and found to be in the PGM 1 + category. His old blood sample hadn't been preserved.

It turned out that he was also unalibied for the after-

noon that Dawn Ashworth had been murdered, but he
died of natural causes just hours after being questioned
by members of the murder squad. He'd been placed in
the top category of suspects and they'd spent months
trying to get court approval to exhume his body.

Dracula jokes were rampant. Nobody was safe from
these vampires: neither the living *nor* the dead.

The Ashworths had decided to take a trip that summer of 1987 to visit Robin's sister who'd emigrated to
Australia seven years earlier with her son. Robin and
Barbara planned their departure carefully. They left on
Thursday, July 30th, and arrived in Sydney on Saturday
morning, August 1st. By crossing the international date-
line they'd managed to make July 31st *disappear*. That
dreaded anniversary of their daughter's death just didn't
happen.

# 25
# Unguarded Moment

His emotional reactions are simple and animal-like, occurring only with immediate frustrations and discomfort. However, he is able to *simulate* emotional reactions and affectional attachments when it will help him to obtain what he wants from others. . . . His social and sexual relations with others are superficial but demanding and manipulative.

The simple psychopath's main characteristic is an inability to delay the gratification and biological needs, no matter what the future consequences to himself or to others.

—ROBERT D. HARE, *Psychopathy: Theory and Research*

As the summer of 1987 began to burn itself out, the murder squad had some of their most difficult times. They drove blood buses to housing estates and factories in order to call people out. In larger work places they even took a doctor with them: a daunting display of mobile blooding. But they were exhausting their bloodlust.

They tried other tacks. They raided a traveling fair in Blaby with two dozen officers, searching the caravans of carnival workers. And they caught a flasher on a village footpath, a professional tennis player who was a psychiatric patient at Carlton Hayes. But he was good only for a few lame jokes about flashing and tennis balls. Always they returned to blooding for the answer.

The DI's, Pearce and Thomas, often went to the blooding. Those were long nights when they bloodied, and sometimes the doctors treated them to dinner. The

DI's had to keep it lighthearted for nervous donors as well as weary cops. One night they conducted a lottery where everyone tossed in fifty pence and guessed how many they'd bloody by evening's end. Some of the frightened donors, many of whom had never been in contact with police before, wanted in.

Then one of them said, "Wait a minute! If I win, how will I know?"

"We'll drop the money in your letter box," Pearce told him. "If you can't trust us, who *can* you trust?"

"Okay, I'll have a go!" he said.

Then they planned a prank in which one of the local bobbies, himself scheduled for a blooding, was to pose as a civilian and come in protesting furiously, where-upon four of them were to pounce on him, snap on the handcuffs and carry him to whichever doctor looked most horrified. Supt. Tony Painter got wind of it and stopped that one.

There was a traveling construction worker from Nottinghamshire whom they particularly wanted to bloody, but he was a fugitive on an assault charge and kept avoiding them. The best they could do after much effort was to leave a message for him to ring the incident room.

He complied, demanding to speak to a superior officer. Pearce handled the telephone call, and after a long conversation they struck a bargain. The fugitive agreed to be bloodied if Pearce would give his word of honor not to arrest him on the warrant.

Not only did the fugitive show up on schedule, he brought with him another traveling worker they'd been seeking. Both men were bloodied, and when they were finished and walking out the door, Pearce suddenly appeared and yelled, "Hold on! You can't just walk out!"

The fugitive crouched, ready to run or fight, or both, but he didn't know about the inspector's offbeat sense of humor.

Pearce grinned and said, "Fancy a pint or two?"

He took both men to a pub and stood them some drinks, after which it was discovered they didn't have bus fare. Pearce had to give them five pounds to get back home.

It was like that: trying to keep everybody interested, entertained and, above all, dedicated. Pearce's own dedication had gotten a boost just before the Ashworths left on their Australian holiday. When he'd taken a can of soda to gardener/cop Phil Beeken at the cemetery, he'd found Barbara Ashworth tending Dawn's grave. Pearce had met her on only one other occasion, but Barbara talked to him in the graveyard for thirty minutes.

When Pearce returned to the incident room he commented that whenever someone's killed you *always* hear that the victim was a nice person, but in this case it was true. "Lynda and Dawn were lovely, bubbly girls," he said to his detectives. "Pleasant, helpful, and *ever* so well liked, weren't they?"

He really didn't have to arouse any member of the small group that was left. The hunt for the footpath killer had consumed them all. They were becoming more fearful of the rumors that they were going to be closed down.

The squad held a meeting where everyone put forth arguments to be taken to Chief Supt. David Baker and beyond. They wanted it noted that Dawn Ashworth II had been opened on a restricted budget because the first Dawn Ashworth inquiry had eaten up so much of the budgetary allowance. They pointed out that the reopening should have been treated as a *new* murder inquiry and budgeted accordingly. They promised not to drag in donors so indiscriminately, but said that in the long run it was still cheaper than a time-consuming verification of alibis. It wasn't their fault, they argued, that the laboratory was months behind in analyzing the blood, perhaps even the blood of the murderer, for all they knew.

They'd begun getting time-and-a-third pay for working more than eight hours in a day, as well as £5.54 for a meal allowance. They offered to give it up, as long as the inquiry was kept open.

They began a renewed search into computer printouts of everyone in Britain who'd been imprisoned in the interim between the murders of Lynda Mann and Dawn Ashworth. It seemed a long time between murders for a serial sex killer, at least according to the psychiatric profile.

There had always been speculation that the kitchen porter, who seemed to know too much, could have had something to do with Dawn Ashworth's murder after all. Several of the police continued to believe that it had been *his* motorbike seen parked under the motorway bridge. There were bizarre theories about why samples taken from the vaginal and anal cavities had not shown a transfer of fluid back and forth, the implication being that perhaps two men had raped Dawn Ashworth, front and back, with only one leaving a sample. Perhaps one of them was a voyeur who had assaulted the dead body *after* the murderer was gone! There were macabre theories like that, because that's the way a murder cop's mind works after he's been in the business awhile.

Each of the sixteen officers still on the inquiry reiterated that morale was high, and that there was no doubt they'd flush him out sooner or later, one way or the other. They debated as to how the footpath phantom might try, or perhaps had *already* tried, to beat their system. The consensus was that he would induce a brother or close relative to take the test for him. A few thought he might be gambler enough to take it himself and hope that Jeffreys's system was not foolproof, and who among them could say it was?

Sgt. Mick Mason, like Insp. Mick Thomas, had been on the Lynda Mann inquiry as well as Dawn Ashworth I and II. Only the "two Micks," DC John Reid and Detective Policewoman Tracy Hitchcox had been on all three. Tracy Hitchcox worked with DC Roger Lattimore, who lived in the village and harbored personal fears for his own teenage daughter. Lattimore never forgot to ring the Ashworths or to stop by with hopeful reports as the hopeless months dragged on.

Mick Mason was the CID opposite of Derek Pearce. Where Pearce was fiery, the kind to shoot from the hip (sometimes hitting his own foot), Mason was deliberate, methodical, with a completeness compulsion. He didn't just dot his *i*'s and cross his *t*'s. They said he duplicated every bleedin *i* and every ruddy *t*. He was the kind to stress over the menu at a sandwich shop: Swiss or cheddar? Swiss or cheddar? Swiss or bloody cheddar! But when he finally made up his mind he was implacable.

Mick Mason would come to work fifteen minutes before he had to and might stay hours after he could have gone home. He was one of the first that Pearce and Mick Thomas had chosen when Supt. Tony Painter wanted a squad on Dawn Ashworth II "to sort out the business once and for all."

Until you got to know the big middle-aged cop, he was the last you would imagine in a pub after an evening of blooding—after they got the music going and had a few pints—doing his version of Tom Jones doing "Delilah," complete with bumps and grinds! Mick Mason, "the pub singer," had that other side. But he'd been devoted to Kath Eastwood from the day her daughter had been murdered, and always promised her that he'd never forget Lynda, that they'd *get* the killer. The pub singer was, by his own admission, obsessed with this murder hunt. Possibly, he wanted the killer more than any of the rest of them.

Mason had become convinced that the motorway runner heading toward Whetstone was their man. His fixation on Whetstone was at first subtle, and later not so subtle. He kept finding reasons for going to Whetstone. He usually sought permission from Derek Pearce who was more likely to approve questionable blooding.

"I've been for a walk by the motorway," he said to Pearce one afternoon. "Do you know there's a footpath up to Whetstone?"

Another time he said, "I stopped this chap walking back toward Narborough from Whetstone. He looked like the punk from the Lynda Mann enquiry."

After several of these, Pearce finally said, "This is one you're dragging in on your own private sweep of Whetstone, isn't it?"

The murder squad had arbitrarily concluded that a blooding cost about thirty pounds. Pearce finally got to the point where he'd say to Mick Mason about a Whetstone man, "Well, is he worth thirty quid?" which would cause Mason to grin and disappear with blood in his eye.

Even with Pearce's "when in doubt, bloody him" philosophy, it got a bit much when Mick Mason began tying up the computer with descriptions of punkers, wanting print-outs on suspects with a residence in Whetstone.

When frustrated voices were raised in the incident room, when the possibility of closure loomed, nobody even looked up if it was Derek Pearce's voice; they were used to that. But when Mick Thomas started raising his voice, as one later put it, "We'd think, 'Blimey! Maybe something *is* wrong!' "

The three-month duty charts had been changed to one-month duty charts. As far as the top brass was concerned, the end was near, and *that* was obvious to the two inspectors. The remaining sixteen held a very tense meeting with Supt. Tony Painter. He informed them that Chief Supt. David Baker was getting great pressure from the chief constable who in turn was being pressured by the Home Office. The inquiry could not stay open indefinitely.

There was an extraordinary clamor at that meeting. People wondered aloud what the press would make of a surrender after four years of hunting the footpath killer. Sgt. Mick Mason openly suggested they should have the courage to begin blooding other places. Like Whetstone, for instance.

Tony Painter became annoyed. He said, "You will not mention Whetstone. We will not bloody Whetstone!" Of course, he didn't know that Mick Mason was *already* blooding Whetstone.

Derek Pearce jumped in to say, "All right, let's pack it up and go home!"

Painter rebuked Pearce about the need for a DI to control himself, but the clamor persisted. Somebody actually said that if the inquiry was closed, the Police Complaints Board should bring a complaint against the chief constable himself!

Baker and Painter and their superiors were facing kamikaze dedication here. Maybe they realized that these last sixteen were foundering in a bloodlust frenzy. They might bloody every goddamn mammal in Leicestershire!

* * *

A new television story was aired that didn't exude confidence. Chief Supt. David Baker, Supt. Tony Painter, DI's Derek Pearce and Mick Thomas were all video-taped by a news team during a blooding session. Baker made another appeal. He said, "We have not got that vital piece of information which allows us to put the jigsaw together completely."

When he was finished, newsmen made sotto voce comments about whether or not the squad had *any* puzzle pieces. The announcer called Baker's statement "a painful admission."

More painful to the murder squad was a visit by an inspection team from the deputy chief constable and the high sheriff of Leicester, who, after being given a brief summary of the mountain of work accomplished by the inquiry, had only one comment: The sign they'd posted for civilians that said, COFFEE 10P, TEA 5P, was "unprofessional."

Such is the policeman's lot, as Gilbert and Sullivan had long ago observed.

On a more upbeat note, a Midlands newscaster said, "As more men come forward the net slowly closes on the killer. If the police hunch is right, and he *is* a local man, he dare not run the risk of giving blood."

On the day of that newscast, David Baker offered a statement to the print media—a prayer almost—that proved to be prophetic. He said, "Somebody's *bound* to say something in an unguarded moment. Now *that's* the kind of information we need!"

The beginning of an answer to Baker's prayer had already taken place on the 1st of August, one year after Dawn Ashworth's murder. It happened in a pub.

The Clarendon Pub in Leicester was a pub for locals: students, university people, journalists. It was a bit

Bohemian in an area that had become trendy. A nice pub, the Clarendon had salmon-colored drapes and valances, coordinated wallpaper, and plush banquettes. It was near one of the Hampshires Bakery outlet shops, off Queens Road.

During the lunch break on that Saturday afternoon Ian Kelly went to the Clarendon, along with a twenty-six-year-old woman who managed one of the bakery outlets. Another woman and a young man, both Hampshires employees, tagged along.

They sat in the busy pub having a "cob," a Leicester snack consisting of a roll filled with meat, cheese or anything you fancy. The talk turned to bakery tittle-tattle, centering on Colin Pitchfork, whom the manager of the outlet shop knew by sight and reputation.

They gossiped about Brown Eyes and her stillborn, and the fact that Colin couldn't stay away from women. As Ian Kelly sipped his drink, a bemused smile crossed his face and he blurted, "Colin had me take that blood test for him."

The bakery manager said, "*What* test?"

"For the murder inquiry?" the male companion asked. "That one, Ian?"

Ian Kelly got up and went to the bar for another pint. When he was gone the bakery manager turned to the other young man and said, "What's that all about?"

"It's odd," the young baker said. "Colin asked *me* to do it too. Offered me two hundred quid to take the blood test. He's just scared of coppers. A weird bloke, that Colin."

The shop manager was deeply disturbed. She tried to broach another question, but it was lightly dismissed as though the implication was preposterous.

Still she couldn't get it off her mind. A week passed, and she took aside the young baker who'd been offered

the money and said, "What are we going to *do* about
Colin Pitchfork?"

He said, "Leave it. He's a friend. You don't even
know him."

She *couldn't* leave it, but she was fearful of involving
someone in a double murder—someone who might be
innocent—not to mention getting Ian Kelly into police
trouble.

Three days later while the bakery manager stewed,
history was made in London at the Old Bailey. Genetic
fingerprinting was used in a criminal court for the first
time in the case of a man accused of unlawful intercourse with a fourteen-year-old mentally handicapped
girl who'd given birth to his baby.

Dr. Alec Jeffreys was quoted as saying, "The use of
the test in a court case is exciting for us. It is an historic
occasion."

The bakery manager knew that the owner of the
Clarendon Pub had a son who was a police constable.
She inquired but found that the bobby was on holiday.
It was six weeks before she rang him up.

It had been a good summer for Carole Pitchfork.
She'd been noticing a marked improvement in her husband's attitude since she had allowed him to return
home in March. She felt that he was trying very hard to
make a go of their marriage. He seemed to be maturing
and accepting responsibility for his past actions. She
didn't even have to nag him to change clothes anymore.
He was dressing better, as befit a budding entrepreneur.

His scheme for opening the cake-decorating studio
was beginning to jell at last. Colin had been to a banker,
and was discussing things like cash flow with an accoun-

tant. He'd even accepted a small commission to make a birthday cake for a policeman's twenty-first birthday. It was cleverly conceived and skillfully executed. The policeman loved it. Colin had done an icing sculpture of a bobby's helmet, alongside a set of steel handcuffs.

Friday, the 18th of September, started off in a river of blood like all the others. Derek Pearce and Gwynne Chambers were on a London run to pick up four blood samples and to interview one man. Whenever they took trips like that they'd call in several times a day. But they got caught in motorway traffic coming back and it was some time before they could get to a phone box. They found one occupied by a girl who had about three pounds, all in tenpence coins, spread out in front of her.

She gave the impatient detectives a glance or two but wasn't about to give up the phone. They jumped back in the CID car and kept going.

When they got to the office at 9:00 P.M. there were messages all over the door. One said, "Don't go home!" Another said, "Got a job on!" A third said, "Don't go home. Got a job on!"

When Pearce got to his desk he found a huge one saying, "DON'T GO HOME!"

Phil Beeken had taken a telephone message that afternoon from a bobby whose father owned a pub near the Queens Road outlet shop of Hampshires Bakery. Beeky relayed the information from that telephone call directly to Insp. Mick Thomas and they pulled an old house-to-house *pro forma* from the Lynda Mann inquiry. They compared the signature of the resident of a semi-detached house in Littlethorpe with the *pro forma* from his blooding in January. The two signatures of Colin Pitchfork didn't match.

Mick Thomas and Phil Beeken tried to keep each other from getting too excited. After all, signatures can change over a period of three years, particularly with young people. But Pitchfork wasn't a kid. Then they looked at each other and decided, The hell with it! They were over the moon and rising!

Mick Mason was telephoned at home and given the job of immediately contacting the others from the bakery who'd been present in the Clarendon Pub when Ian Kelly blurted an admission during an unguarded moment. Thomas and Beeken went to the manager's house and took her written statement.

She began by saying, "This is probably a waste of your time, but my conscience *forced* me to ring the police." She kept apologizing until they reassured her.

By the time Mick Thomas and Phil Beeken hooked up with Derek Pearce and Gwynne Chambers later that evening they were practically hyperventilating.

Mick Thomas said to Pearce, "Roger and Tracy are still in Yorkshire trying to bloody some bloke! You and Gwynne were in London! Everybody else had gone home! I was going crazy with no one to tell!"

One of them noticed something very peculiar. The conversation in the Clarendon Pub, that unguarded moment, had occurred exactly one year after the day that Dawn Ashworth's body lay undetected in a field by Ten Pound Lane. It seemed to be an omen.

They unanimously elected to go immediately to a pub, and they did. While drinking his second pint Pearce said that going to bed was out of the question. He wanted morning to come without having to sleep through the interim. Mick Thomas suggested that they'd better not get drunk because of the importance of the following day. But they didn't have to worry—the booze couldn't compete with the adrenaline rush.

Each man later reported that he spent a near-sleepless night. Each later reported that he felt he was facing the most important day in his police career.

As far as Pearce was concerned: "It was the most important day of our *lives*."

Ian Kelly had not been having an easy time at the bakery since he'd given blood for Colin Pitchfork. It seemed as though too many things were going wrong, and Colin Pitchfork was always around to "help" him. Once when they were making buns, Ian burned them. Colin observed the error and told Ian not to worry, he'd take care of it. Ian later heard that Colin "took care of it" by informing the foreman.

There was a more serious incident when Ian was making buns with another baker. A huge steel machine cover was propped against a wall. Ian pushed a baking trolley past it and was absolutely sure he had sufficient clearance, but somehow the heavy metal cover fell over and crashed into his partner's legs.

The man bellowed and swore and accused Ian of crippling him. It turned into such a row that the gaffer came out and shouted, "Stop behaving like kids, the two of you!"

The injured baker was so outraged he told the boss to stuff it. The baker quit his job that day, saying that Ian Kelly was the one who should've been sacked.

Ian went back to work, absolutely baffled as to how the machine cover could've fallen. Until he later learned that Colin Pitchfork had been standing nearby when it happened. It was *beginning* to look like somebody wanted him out of Hampshires Bakery.

\* \* \*

On the morning of Saturday, September 19th, it was decided that Pearce and Chambers would arrest Ian Kelly. And they might arrest the young baker who'd been offered £200 by Colin Pitchfork, depending on his answers. Mick Thomas and Mick Mason were to call on that young man. Even though they were off duty, DC Brian Fentum and Phil Beeken insisted on being there. Nothing could've kept them away.

Ian Kelly opened his door that morning to a pair of visitors he knew weren't selling magazines. Derek Pearce showed his warrant card and said, "We're from the murder enquiry incident room at Narborough, investigating the murders of Lynda Mann and Dawn Ashworth. Have you given a blood sample regarding those enquiries?"

"No, not me!" Ian said.

"I don't believe you," Pearce said. "I have reason to believe you've given a blood sample."

"No, I haven't!" Ian said.

"We've talked to other people at the bakery," Pearce said. "I believe you *have*."

"Yes, you're right," Ian said. "I did it for another lad at work."

"Who's that?" Gwynne Chambers asked.

"Colin Pitchfork," Ian Kelly answered.

Pearce said, "I'm arresting you for conspiracy to pervert the course of justice and we're taking you to Wigston Police Station."

"Yes," said Ian Kelly. "I'll just put me shoes on."

They took Ian Kelly to the station, which was already humming, and put him into an interview room where his statement was recorded.

Pearce said, "I must tell you, you do not have to say anything unless you wish to do so, but anything you say may be given in evidence. Do you understand that?"

Ian began by saying, "Yes, well, the gentleman in question, Colin Pitchfork, he come up to me and asked if I'd do him a favor. I didn't know it were for them murders. I didn't know what it were really for cause he didn't explain what it were for. He just had to give a thingybob cause he got a letter from the police station."

Then Ian related the story that Colin Pitchfork had told him about giving a sample for the other bloke, and Ian told about the photo strip and altering the passport. But he stuck to his claim that he didn't know that the blooding was for anything as serious as murder.

Derek Pearce didn't look *quite* as dangerous as a Shi'ite with an AK-47 when he said, "Yeah, you're Mister Muggins. And you've just gone along and given the sample. And he got what he wanted: full protection. *You've delayed us eight months!*"

Ian started to understand what was facing him. He said, "Well, when I went to his house, more or less . . . well, the day before, he *told* me it were a murder enquiry. But I didn't know *which* murder it was at the time!"

And he admitted to having been given a little schooling on the dates of birth of the children and other personal information. He said, "I knew it were for a murder but I didn't know whose it were for, cause at the time when I walked in I were *that* sick. I'd got a temperature. I was feeling really low. I mean, when I began writing his signature I got shaking like a leaf!"

Supt. Tony Painter was called in that afternoon and found Derek Pearce bobbing and bouncing like a dinghy in a storm.

"Let's go nick him!" Pearce said to their commander.

"No, take it all down on paper," Painter said. "And *then* go get him."

Pearce said, "We want him *now!*"

"I'm the boss and I say paper first," Painter said.

"Quite right," Pearce said. "Paper first."

So they had to wait another two hours until all statements were transcribed and put in some semblance of order. By the time six of them got to the house in Haybarn Close, the blue Fiat was gone. There was nobody at the Pitchfork home. One stayed; the rest returned to the station, *trying* to be philosophical. After all, they'd waited four years.

# 26
# Blind Terror

The most important feature of the psychopath is his monumental irresponsibility. He knows what the ethical rules are, at least he can repeat them, parrotlike, but they are void of meaning to him. . . .

No one wears the mask of normality in so convincing a fashion. He is strikingly cool and sure of himself in situations where others would tremble with sweat and fear. . . . He retains a superhuman composure.

—PAUL J. STERN, *The Abnormal Person and His World*

A surveillance and arrest of a major felony suspect is done differently in Britain than in the United States. In Britain a suspect under observation is often allowed to enter his house so that he can't run away. In a gun-crazy country like the U.S. the last thing the police want to do is let *any* suspect enter his house, where he may have enough firepower to take the Persian Gulf.

Late that afternoon the murder squad allowed the blue Fiat to pass into Haybarn Close and proceed to the end of the cul-de-sac. They waited until Colin Pitchfork parked the car, until the entire family was safely inside the house.

Derek Pearce, who said he lived to cover back doors, ran around to the rear with Gwynne Chambers. The two Micks, Thomas and Mason, went to the front. Phil Beeken and Brian Fentum backed up the two Micks. At 5:45 P.M. Mick Thomas knocked.

Carole Pitchfork later said, "At first I thought they were insurance men. I thought perhaps it was about the car accident on Narborough Road. They came in and said they were police officers and asked to speak to Colin in private."

Mick Thomas and Mick Mason walked Colin Pitchfork into the kitchen while the others stayed in the living room. Phil Beeken later said, "I saw him and thought, Yeah, it's him! He looks the way our man *ought* to look! It's *him!*"

Mick Thomas said to Colin Pitchfork, "From enquiries we've made we believe you're responsible for the murder of Dawn Ashworth on the thirty-first of July, 1986. We believe another man gave a blood sample for you. I'm arresting you on suspicion of that murder. I must inform you that you don't have to say anything, but anything you say may be taken down and given in evidence. Do you understand?"

Colin Pitchfork very calmly said, "First give me a few minutes to speak to my wife."

Mick Thomas had a *feeling* from the look of resignation on Colin Pitchfork's face, and so did Mick Mason, who suddenly asked, "*Why* Dawn Ashworth?"

Colin Pitchfork replied, "Opportunity. She was there and I was there."

Mick Thomas then asked, "What do you want to speak to your wife about?"

"It's going to be a long time till I see them again. You've got to let me say goodbye."

Just then Colin Pitchfork's four-year-old son cried, "Daddy, the telly won't work!"

Mick Thomas nodded an okay and Colin Pitchfork walked into the living room to adjust the tuning. Mick Mason grabbed all of the kitchen knives off the counter, just in case, and opened the back door. When the pub singer got outside, he did a little saber dance with those knives, and Derek Pearce knew it was over.

A few minutes later, Pearce and Mick Thomas were in the little kitchen with Colin Pitchfork, while a very frightened Carole Pitchfork was asked to take the kids upstairs.

Colin Pitchfork asked, "Why's there a need for the other officers to be going round my house?"

"There's a number of things we need to search for," Mick Thomas told him.

"Like what?"

"The passport that was used."

"It's not here," Colin said. "Honestly. It's at work. Let me speak to my wife."

Mick Thomas said, "You can speak to your wife, but only in my presence." Then he called for Carole Pitchfork and she came down and entered the kitchen.

She looked from one to the other. She looked at *him*, leaning against the kitchen cupboard.

He moved forward and tried to put his arms around her, but she pulled away.

"They've come to arrest me," he said.

"What *for*?" she asked.

"For them murders."

"But you went for the blood test!" she cried. "And you got a letter saying it was negative!"

"I didn't go," he said. "Ian went."

"Did you do it?" Carole Pitchfork asked him then.

He didn't answer.

"Did you do it?" she asked again.

Still he didn't reply.

"Did you?" she asked.

"Yes," he answered.

And she flew at him. Neither detective was ready for *her* to launch an attack. They jumped in between. Mick

Thomas grabbed Carole and bundled her out the door, but not before she directed a punch and kick at Colin Pitchfork. She missed her husband but managed to punch Derek Pearce in the mouth and kick him in the groin.

On the way to the station, in the back of the CID car, Colin Pitchfork said to Mick Thomas and Mick Mason, "I *must* let a few people know what's happened to me before they read it in the papers. Then I'll tell you everything." He paused and said, "But I want to do it my own way. Because it's really a story of my life, not just the story of a month or two."

So he was attempting control even before they got him to the station.

Mick Thomas assigned a man to phone everyone on the team and to keep ringing until they were all found. He didn't want anyone to get the news secondhand, not after *four* long years. He discovered that he and Mick Mason were going to finish this job themselves, even though a superintendent would ordinarily conduct such an important interview. But there was no senior officer present.

Every member of the team was bewildered. Supt. Tony Painter had gone home before they'd made the arrest that would conclude the most important murder inquiry ever conducted by the Leicestershire police. But Tony Painter was a very proud man. Perhaps, after clinging so long to his belief in the guilt of the kitchen porter, he just needed some time to deal with it.

After Colin Pitchfork made his phone calls and had a cup of tea, Mick Mason turned on the tape machine. The prisoner said, "You know, before we actually go into the rigmarole of the details, can we sort out other bits and bobs?"

He began with his earliest memories. He started by telling them of a friend he'd had when he was eleven. He'd never had many friends in his life. He told them of the Scouts, of his triumphs there. When he had finished describing his first fourteen years on earth, they wanted to *please* talk about the murders. But it got him angry. Colin Pitchfork threatened to shut down the confession unless they did it *his* way, beginning with his earliest recollections to the present. Every fascinating event in his fascinating life. It took *hours*. The stage was his. They were bored to tears.

After Carole Pitchfork had attempted to attack her husband, she lost touch with the flow of events.

"Next thing, I was outside with all those coppers," she later recalled. "And soon he was gone from our house."

When she got her wits about her, she asked the police to call a neighbor to look after the kids. When the neighbor took the children home for the night, her older son cried. He wanted to watch *The A-Team*, and to make pictures with the new set of paints they'd just bought him.

Carole remembered shouting at the cops about the searching, and being told, "We have to have your permission to search."

"Get on with it!" she cried. "Get *on* with it." Then she picked up a toy fire engine and threw it across the room.

After she deliberately pushed over the bookcase, Derek Pearce persuaded her to calm herself. They searched the house until 1:00 A.M. but found no significant evidence.

The couple who lived next door came to assist her, and along with Carole they consumed a bottle and a half of brandy during the course of the police search. Then Colin's brother arrived with a girlfriend and offered to sit up all night with Carole. She rang her father who cut short his vacation to run to his daughter.

Carole got a blinding headache that evening, the worst of her life. Pearce offered to send for a police surgeon, but Carole refused. Then he rang her family doctor who came to the house, examined her briefly, and said, "You've got a migraine."

When Carole said, "I don't get migraines," the doctor replied, "You've got one now."

The journalists came the next morning. In droves. She hid inside behind locked doors and drawn curtains. The reporters photographed everything in sight—the house, car, street, windows—waiting for those curtains to move or even twitch. Their cameras rooted into every crevice.

One of Carole's more irrepressible neighbors, who finally despaired of making the reporters go away, went to the window, opened the curtains, raised her T-shirt and bared her breasts.

"Put *those* in your fucking paper!" she yelled.

They were the only things they didn't photograph.

When at last Mick Thomas and Mick Mason were *permitted* by Colin Pitchfork to ask questions about the murder of Lynda Mann, the prisoner again took his time setting the stage. He described how he and Carole had been preparing to move into Littlethorpe in December, 1983, and how he was recording music that night for a going-away party. He told of dropping off Carole at the college and going on a wander for a girl to flash, then of driving down Narborough Road and turning on Forest Road where he saw the young girl walking.

"Which way was she walking?" Mick Thomas asked.

Colin Pitchfork grinned triumphantly and said, "This is one of the questions you've *always* wanted to know, isn't it? She was walking from Narborough up to Enderby. At that time the new housing estate wasn't there, was it?"

"And then?"

"Then I turned the car around, my red Ford Escort. I reversed in the drive of The Woodlands, there across from the mental hospital. And I left the car in the drive. The baby were in a carrycot in the back. Always been a big believer in restraints, you know. Car restraints."

Apparently satisfied that he'd vindicated his parenting, he said, "I set myself a walking pace to meet her under the light. It was very dark there. And cold. With me only wearing jeans and a jumper. Have to be under a light. It's no good flashing yourself in the dark, is it? And when she got up to where I stood, I did it. The shock, I would say *shock*, made her run backwards toward the footpath. She left the main road."

"You were surprised?"

"You see, the way she'd been traveling toward where I parked, I couldn't flash her and run back to my car right away or she'd have seen it. If she'd just walked by me like all the others did, I would've started walking down the road, then doubled back and got the car after she got out of sight. But it never happened like that."

"And then what?" Mick Thomas asked. "After she ran in shock toward the footpath?"

"It were the *thing*," he said. "The flashing. It got the excitement. . . . It was *there*. She suddenly ran herself back into a dark footpath. On her own. There were an open field by the footpath. She had run herself into . . ."

Mick Thomas said, "A worse position."

"Yeah. She ran herself into a dark footpath on her own, and she just *froze*."

"And what did you do?"

"I went up to her and grabbed her and she didn't really resist me when I grabbed her. I took her off the footpath and had a conversation with her."

"What did Lynda say?" Mick Mason interjected.

" 'What are you doing to me? What about your wife? Where have you come from? What are you doing this for? What have I done to you?' " Then Colin Pitchfork said to the detectives, "This is the thing I don't understand about flashing. One percent of the time you get someone who goes mad and screams and you have to disappear quick. But all the others walk by you. Just walk *by* you and ignore you. But *she* turned and ran into a dark footpath. She backed *herself* into a corner."

"*Her* mistake?"

"If she'd walked by, the situation would've disappeared. But she ran back and stood there. She froze. Her two big mistakes were running into the footpath and saying, 'What about your wife?' She'd seen my wedding ring."

"What happened next?"

"By then the urge hadn't subsided at all," Colin Pitchfork said. "It was just getting *stronger*. Because not only had she got *herself* into the situation, she hadn't screamed. She hadn't struggled. If she'd screamed she would've probably scared me off. I *suppose* you'd say I raped her. You'd *have* to say I raped her."

Mick Thomas said, "*You* tell us what happened."

"I raped her in a way," Colin Pitchfork said. "But it wasn't forcible, like if I ripped her clothes off and jumped on her and beat her up. I just said I was going to do it to her."

"Did you remove any of her clothing?"

"No, she did. You might think I'm a bloody crank confessing to it. I know this is a daft thing to say but I can remember every bloody detail because it's haunted me. She had on a black donkey jacket. Her trousers had got zips. She were getting most anxious taking her trousers off because the zip jammed. So I just told her I'd do it. 'Don't you do it,' she said. 'You'll rip my new trousers.' I said, 'Where do you live?' She said, 'There.' I said, 'Where have you been?' She said, 'At my friend's to get some records.' She was trying to calm me down. Talking me out of it."

And at this point, Sgt. Mick Mason—who later admitted his "emotional involvement"—wasn't merely trying to establish the legal elements of rape when he said, "And was she *terrified*?"

Colin Pitchfork just shrugged and said, "Yeah. But rather than scream and struggle and fight, she decided just to let me do it. It were then that I actually satisfied myself. But I suddenly realized that I got myself into deeper shit than I ever got myself into. Before that, if the police had a look at me it was as a flasher. It was a pain in the arse, but never actually did nowt to me, that flashing trouble. But not *now*."

"This was different."

"The thing that was preying on me mind was that she said, 'What about your wife?' She knew I was married. Was that because I'd said something? No! She saw the ring!"

"Your wedding ring," Mick Thomas said.

"Yeah. I also realized I'd got an earring on. And I'd been losing me bloody hair. She could describe those things. She'd almost stopped fighting then. She thought it was over."

"Yes. And then?"

"I suddenly realized I were going to come and live

there in the village, in a month. *She* lived there. Almost certainly she'd see me in the village. The earring, the hairline, the wedding ring. There was no way out. I was trapped."

When they changed tapes Mick Thomas tried to establish exactly where Lynda Mann had been murdered. At first Colin Pitchfork had said on the path, but then he described the killing ground as "a copse," an area beside the path. He didn't seem comfortable with admitting that Lynda had been dragged off the path, through the gate and into the copse, which would have presupposed initial force and violence, such as a hard blow that could have caused the bruise on her chin. Colin Pitchfork said that the gate was already open when she ran into the wooded area beside The Black Pad.

When it was time to talk about the semen having been found on the genital hair, he brushed off the suggestion of premature ejaculation.

Mick Thomas asked, "Were you erect? Was your penis hard?"

"Yeah," he answered at once.

"Did you insert your penis into her vagina fully?"

"Yeah."

"Did you ejaculate inside her?"

"Yeah."

"Fully?"

"I dunno," he said, testily. "I says to her, 'Is it hurting you?' She says, 'A bit.' I says, 'Try and relax. I'll try not to hurt you. Are you all right?' "

"When you say you ejaculated inside her, what happened immediately after you ejaculated?"

"I thought, 'You just can't leave her. Because if you leave her you're going to court.' That's when I strangled her."

"How did you strangle her?"

"Hands. I was still inside when I thought, 'Shit! You've got to do the lot!' I was still inside her when I put my hands up to her throat. Her immediate reaction was to struggle, so I came out of her. I got to kneeling, but she'd twisted around and half sat up. She was struggling."

Mick Mason jumped in then and said, "Do you think she knew you were going to kill her with your hands going to her throat?"

It may have been the look in the older detective's eyes; it may have been the tone of his voice, his inability to maintain the detachment of his younger superior officer. Colin Pitchfork was obviously put off by Mason. He said, "I don't know. I suppose that's difficult to say: what somebody's thinking in *blind terror*."

Mick Thomas, wanting to get the prisoner relaxed again, said, "Yes. And then what did she do?"

"She squirmed sideways," Colin Pitchfork said. "Kicked her arms and legs, jumped half up, started going mad. And then she got sitting up and I leaned me weight on her to knock her down again. I put her on her back, basically to stop her fighting."

Mick Thomas said, "When she was jumping and panicking, did it change your attitude toward her?"

"Yeah, because it became a threat. When she were still calm, I had *control* of the situation."

Mick Thomas said, "When you didn't have control, did you hit her?"

"Hit her? What, physically? I don't think so. I may've knocked her, to get her into a position to strangle her. I never really hit her. Not with any force."

Colin Pitchfork had been trying to explain control and the loss of it while he was trying to control the interview, even as Mick Mason was losing some control and causing Colin Pitchfork discomfort. Unlike other clever sociopaths, Colin Pitchfork had never taken the opportunity to *imitate* appropriate responses to emotions he couldn't feel, therefore his frowns and smiles were utterly out of sync with the subject.

Mason pointed to Colin Pitchfork's bulky upper body and said, "You'd be intimidating even to a big bloke. I'll be dead straight with you. I can't *comprehend* what these girls must have felt!"

Colin Pitchfork stared into the eyes of the big detective, eyes as blue as wisteria, and said warily, "Yeah, I can see by your face that you're *amazed* by it." Then he turned to Mick Thomas for empathy, and said, "I mean, I ain't *proud*. I'm just staying calm because you want the story out of me. So if we do it *together*, easy. Okay?"

If the detectives were looking for frenzy, they'd not find it. The murder squad would forever describe the confession of Colin Pitchfork as "cold," as in "a cold confession of evil." But an outside observer might conclude that the taped confession of Colin Pitchfork revealed nothing more and nothing less than the typically dispassionate narrative style of a self-absorbed, remorseless sociopath. One who, by clinical definition, *cannot* infuse a confession with emotions that he has never felt. Colin Pitchfork and Mick Mason were speaking to each other in different tongues.

When Mick Thomas again took charge, he said, "Yes, and after she struggled, what happened?"

"I got onto the carotid arteries and she was unconscious in a matter of seconds. I would say she was unconscious in twenty or thirty seconds because I put so much force into the carotid arteries. So the oxygen to her brain ceased almost instantly. She had only seconds but she still struggled a bit. I held her for a minute or so, then her body took a natural kind of reaction. A breath, because, I mean, you probably ain't seen a body die, as such."

"And then?"

"I grabbed her coat and pulled her toward a bush eight feet away. By the lapels, pulling backward. There were no point trying to hide the body. Then I noticed she had a scarf on. All I did was tighten that."

"To make sure?"

"Yeah, I seen the scarf and I just wrapped it round her neck and gave it a pull. What you'd call an insurance policy."

"Then?"

"I went back to the car where the baby were still asleep. I had to hurry home and get the taping done because Carole would say, 'What you been doing?' I had to work fast and furious to get the records moving on and off the record player and onto the cassettes. By the time I went to pick Carole up I'd had a wash and a shave."

"After Lynda Mann, did you continue flashing?"

"After Lynda, I stopped flashing for six or eight months," he said.

When a uniformed constable entered the interview room with some tea and a message for Mick Thomas, Colin Pitchfork turned to the bobby and began relating the events of his life to *him*. He seemed disappointed when the constable left. He wanted the attention of *everyone*.

# 27
# The Cake

No sense of conscience, guilt, or remorse is present. Harmful acts are committed without discomfort or shame. Though the psychopath, after being caught or confronted with a brutal act, may verbalize regret, he typically does not display true remorse.

—RIMM and SOMERVILL

Colin Pitchfork's next interview was about Dawn Ashworth. He began by saying, "I was riding the Honda Seventy when I seen the girl enter the footpath. I was out to get food coloring for a cake. I parked the bike and put me hat on the handleclip and just walked after her into Ten Pound Lane."

"Was there anybody around?"

"Nobody. Nobody ever saw me. They saw lots of other people, I guess, but not me. There I was in broad daylight, wearing jeans and a jumper and a bottle-green nylon parka jacket."

"Go on."

"When I were following behind Dawn I had this gut feeling. It was saying, No no no no no! But the other side of me was saying, Just flash her. You've got a footpath. You've got all the time in the world. Even if she runs off screaming no one will ever see you. No one will ever know! Who's going to know?"

"You were following behind?"

"Yeah, but I had a hard time catching her. She

341

walked fast. I finally jogged past and turned and half smiled, as if to say hello. I tried to get ahead. I tried to get set, but she was on top of me. I didn't even have time to open me bloody trousers. There's rules to how I play that game. I prefer to do it in a way that satisfies me. I still had me motorbike coat done up and I was a bit out of breath. She had plenty of room to walk by me. I had got to this point on the path, by this little opening. This gate. You got to make a decision whether you're going to do it or not. So I turned and walked back toward her and exposed myself. She didn't say nothing."

It was Mick Mason who then asked, "Was your penis hard?"

"Don't know," he said. "Can't remember."

Mick Mason said, "How clear is this in your mind? The whole thing? The murder?"

"Crystal," Colin Pitchfork said.

Mick Mason said, "Then what?"

"They always have room," Colin Pitchfork explained. "No matter where I were exposing meself. No matter where. They always have room to walk by me. It's the easiest way. You shock them. They walk by you and then you got your exit route clear, and go where they come from. If she'd have ran back down the path she'd have blocked my exit route. If she'd have ran back screaming on Narborough Road I'd have had trouble getting back to the bike. So I had to stand to one side of the path or I'd have blocked her way and *my* escape route."

"So what did you do next?"

"Now I know it sounds very familiar," he said, "but she jumped *into* the gateway."

"Like Lynda?"

"Yeah! I just moved forward and pushed her toward

the gateway. It was the same as Lynda! You're there
again! It was the same, but like, worse. And I thought,
No no, don't touch her. Leave her alone. But then
once I grabbed her you're in a situation that if a girl in
Narborough gets grabbed, they'll immediately go to the
Lynda Mann enquiry, won't they?"

"Continue."

"Well, I still got me wedding ring on. I thought,
Shit! I knew I were heading into the same thing. It was
a reoccurrence. I had her just round the shoulders and
round the mouth when she squealed. I had her from
behind cause she turned her back when she jumped
toward the gate. I very nearly let go of her but the one
side got the better of me. See, I still had the motorbike
jacket on. She'd report me and I can't take it off and
just start riding the motorbike without a motorbike
jacket on. I moved toward her and she screamed. A
loud scream."

"What did you do?"

"As I put my hand on her mouth and half leaned on
her, the gate opened on its own. I pushed her into the
field and she never said much at all. She submitted
more then. I would say even more than Lynda."

"The gate opened on its *own*?"

"We both went into the field through the gate. . . .
It weren't on a catch."

"What did she say?"

"She kept saying, 'Please don't do it. Please don't!'
But she never actually fought or anything. Before I got
her down, before I actually raped her, I knew there was
no way I could go back. I could only go forward. The
same feelings were coming back. That I was in a trap
again."

"What was said?"

" 'Shut up the bloody shouting!' I said. I pushed her

away from the gate and she fell over. That was really
the first time I'd thought of doing anything to her."

"Just at that moment, not before?"

"You find yourself in an open field. Nobody had
actually come to her rescue or anything. There weren't
even a chap walking a dog. When she fell she just got
quiet, looking at me. And more than anything I were
watching the gate. Just to make sure no one was coming. I think that's when I began to get the . . . I guess
the only way you can describe it is screaming voices in
your head. And it seemed like ages but I'm sure it was
only seconds. Like, I don't want to leave her. Don't
want to go away. Do it again here! But saying, No no no
no! And the other side was saying, She's here. She's a
young girl. She's laying down in a bloody field in the
middle of nowhere. There's nothing easier! It were a
confliction."

"And what happened?"

"She said, 'You're not going to hurt me, are you?'
She babbled something and said, 'You're not going to
rape me, are you? I'm a good girl. I go to church.'"

Mick Mason asked, "And what was *your* reaction to
that?"

"I said, 'Oh, bloody well shut up!' I reached under
her back and just pulled her knickers straight off. Once
I got them off she half picked them up, then I just
rolled on top of her."

"And you raped her?"

"I did it, and got off. She were very calm. She sat up
and she said, 'Have you finished? Can I go now? I won't
tell anybody! Please, I won't tell anybody! Honest! Just
go and leave me alone! Please!'"

Mick Mason again had to ask it. "And was she
*petrified*?"

"Yes," Colin Pitchfork said. "But it was *calm* petrified. 'I won't tell anybody! Just leave me alone! Just go and leave me alone!' "

Mick Mason said, "Right. So what was going on in *your* mind at this stage?"

"Screaming to meself, Shit! You done it now! Not only was the problem that you raped her. But you've already got one murder on your hands. After you murder once and murder again you got a better chance of being caught, but . . ." Then Colin Pitchfork showed Mick Thomas a grin of familiarity and said, "But like *you* said, Mick, the sentences are not like in America. Two murders are okay. It's not *twice* the sentence for two."

Mick Mason may have been getting close to his limit. He said, "These voices shouting to you are not going on about prison sentences, are they? They're not *really* voices, are they? It's your innermost thoughts, your *conscience* telling you what you can do and can't do, isn't it? I mean, surely, your conscience is telling you, *Don't!*"

Colin Pitchfork, who'd spent years telling probation officers and psychiatrists what they wanted to hear, said, "Yeah, my conscience is telling me, Don't. But at the same time it's saying, She'll identify you!"

Mick Mason said, "Okay, so in view of your conscience telling you things, what did you do next?"

But actually, Colin Pitchfork had *never* mentioned the word "conscience," nor even described one, not in any of the interviews he was to give, except when Mick Mason forced the issue.

The prisoner said, "I left her for a second and let her sit up."

"What did she do and what did you do?"

Narratively leaping from first to second person, and past to present tense, Colin Pitchfork said, "She sat up.

And I had got to kill her. You can cover your tracks. You can get away with it if you kill her. She had her back to me after she sat up. Which presented me with the ideal opportunity to do a strangle hold. To get her from behind."

"With your hands?"

"No, with a judo strangle. Different than the simple thumbs on either carotid."

Colin Pitchfork then exerted his power and control. After all, this was *his* life story. Instead of choosing to demonstrate on the "good" detective, he chose the "bad" one, the one who disapproved of him. The prisoner got up and moved around behind Mick Mason and put his arm around Mason's shoulders and his forearm across his throat. And clasping his right hand across his left elbow, he flexed his left arm, putting pressure on either side of Mason's throat with the biceps and forearm.

"Yes, you can see it, can't you?" Colin Pitchfork said. "It's instant."

Then he sat back down and found it very hard to repress a certain amount of pride. "In fact," he said, "I learned judo at the Caterpillar Judo Club at Desford."

Mick Thomas said, "After raping her did you do anything else sexually to her?"

"No," he answered quickly.

"Did you insert yourself into her bottom?"

"No," he said.

Mick Mason said, "At all?"

"No!" he said.

"Carry on, then."

"Her jacket had come off when I applied the strangle hold round her shoulder. I started to take a ten-pound note out of her purse to make it look like robbery. Then I thought no, I couldn't see no point in it. I thought, what with money being tight at home, Carole might

find it and say where did I get ten quid from? The only
cash we had was used for groceries and petrol."

"So you didn't take it."

"Then I realized lots of people walked their dogs on
the path, so I moved her body in toward the stinging
nettles. I saw a six-foot log. I half dragged her body into
the stingers and hedge. I covered her with the log and
threw her jacket down the hedgerow. Then I had a bit
of panic."

"Why?"

"I lost me watch. Which I don't know if you ever
found, did you?"

They had not, but Mick Mason wanted to get back to
the dying girl. He said, "What happened when you put
the strangle hold on her? Specifically."

Colin Pitchfork looked at the older detective and
said, "You'll appreciate how quick it killed her if I just
show you the effectiveness. Can I just do it again? On
*you?*"

Mick Mason said, "No. What was the effect on the
girl? How long did it take?"

"Seconds," Colin Pitchfork said, again annoyed with
Mason.

Mick Thomas asked, "Did she say anything?"

"No. She just *died.*"

And then, like any reasonably adept sociopath who is
confronted with the conscience of others—something
he considers a *weakness*—Colin Pitchfork said, "I know
I'm talking about it coldheartedly. I don't *feel* that way.
She died a damn sight quicker than Lynda Mann. Because with Lynda Mann I didn't go straight for the
carotid artery, cutting off blood to the brain. I can tell
you, this is a recognized Japanese way of getting somebody. I been instructed in judo, working on a mat."

By then, Sgt. Mick Mason wasn't interested in Colin

Pitchfork's prowess in judo, or scouting, or cake baking. Mason was getting very near to the *end*. He said, "The rape was *very* traumatic. The girl was ripped to bits around the vagina and bottom. Can you explain that at all? She was absolutely ripped to *bits*."

"No, I can't," Colin Pitchfork said.

Mick Thomas said, "Did you realize that the girl was a virgin?"

"She told me halfway through the rape," Colin Pitchfork said.

"Did you have any blood on you?"

"No."

Mick Mason started firing multiple questions. He said, "Did you have any control over what you were doing during this rape? Were you completely carried away? And what *were* you doing to this girl? Exactly."

"I wouldn't say *control* over what I was doing," Colin Pitchfork said. "But I'd say I can *relive* it second by second."

Mick Thomas said, "The medical evidence shows that she was entered per anus as well."

Mick Mason said, "Do you understand what *that* means?"

Colin Pitchfork was *really* annoyed now. He said, "Yeah, I know what *that* means."

Mick Mason started getting rhetorical. He said, "You went up her bottom. That's shown because tissues were ripped inside her. Now, we haven't thoroughly covered that particular aspect of the murder. It may be that you're aware it happened, but it's too distasteful to discuss. But it's a point that we feel we should clarify with you. To discuss your mental attitude at *that* moment. That you were in control. The answer you gave suggests that you knew exactly what you were doing, albeit you were going along this natural progression. . . ."

The prisoner interrupted him, saying, "I wouldn't say that I *knew* what I was doing. Although I can recall it, it wasn't as though I had the *control* to stop it."

Mick Thomas said, "I'm saying that you were under control in that you knew what you were doing, but couldn't *stop* what you were doing. Now if I've read you right, you would *know* if you inserted your penis in that girl's bottom. Whether because you chose to do it or whether it happened because she was struggling, or you were excited, or whatever." He tried to placate him by saying, "I mean, nowadays it's perhaps not as unacceptable as it was twenty years ago. Probably consenting adults *do* partake in that particular way."

But Colin Pitchfork wasn't buying consenting adults. He said, "As far as I'm concerned, I rolled on top of her and raped her and that's it as far as I'm concerned!"

Mick Mason simply could not let go of his need for a sign of contrition. He felt this was the most "evil and chilling" man he'd ever met or ever would meet in his police career. He was trying to do what cannot be done: locate a nugget of genuine remorse in a sociopath. He said, "She *must* have been in traumatic pain when you were doing this."

That simply irritated Colin Pitchfork all the more and he said, "So what's *that* mean?"

Mason said, "It means that the girl as you describe her, and as she's been described *to* you, is this straightforward girl. Well, were you using some other means to control the situation but not heighten her trauma too much?"

"Such as?"

"Well, with what force were you having to insert yourself, *and* hold her down?"

"I wasn't! I told you she was already down. She was laying down. She wasn't attempting to get up!"

Fearing that the prisoner's cooperation was being jeopardized, Mick Thomas again made peace. He said, "Mick brought that up because obviously you *had* been hurting her. Now, we're not trying to rub it in, but when you think about how you've hurt somebody it's reason for you to feel very uncomfortable about telling it. As you know, a woman's first experience is not the greatest in her life."

Colin Pitchfork said sullenly, "I *asked* her if it hurt and she said, 'Yes, it hurts me!' And I said, 'Just lay still and it'll be done quicker.' "

They tried one final time to get a description of the "horrific" vaginal and anal assault described at the postmortem.

Mick Thomas said, "When the girl's body was found, it was examined by a pathologist and he was able to say that the girl sustained injuries not consistent with ordinary rape. Which suggests *other* violence had been used towards her. Can you explain that?"

"No. Can you be more specific as to what the pathologist actually said?"

"At the moment I haven't got the statement here, but what I can say to you is that even though injuries occurred just prior to death, bruising started. And he was able to say that violence was used, other than that connected with more ordinary intercourse."

"I can't really recall any other violence. I say the most violent bit was when she turned her face and I sort of grabbed her round the neck. It's possible that she may've moved my arm across her neck if there was bruising on the side of the face. I can't remember no overwhelming blows, et cetera."

Colin Pitchfork denied concealing the body as completely as they'd found it, recalling only a heap of nettles, a little hay and a log. But since a police photo taken three feet away couldn't even reveal a body, the bizarre theory would persist with some detectives that perhaps the kitchen porter had found the body, tampered with it, and concealed it more thoroughly. Unlike those in genre murder mysteries, confessions are rarely tidy in real life.

After he'd concealed Dawn Ashworth's body, Colin Pitchfork said he'd walked across the motorway footbridge instead of going back down Ten Pound Lane. He'd taken the route that Robin Ashworth had always urged his daughter to take. Before getting to the other side of the motorway Colin Pitchfork had removed his motorcycle jacket so that no one would describe a cyclist walking.

He'd been entertained by the police inquiry. There was the thing about the kitchen porter on the motorbike. Colin Pitchfork had also been riding his motorbike the day he killed Dawn Ashworth. It was an amusing coincidence, but that's all. It wasn't *his* motorbike they were always writing about, the one parked under the motorway. Colin Pitchfork had left his on a side street near King Edward Avenue. And he certainly hadn't climbed up any embankment and run across the bloody motorway. The fact is, *nothing*—not a single lead the police had announced in four years—had *ever* applied to him.

Nobody had ever seen him. It was just that easy to rape and murder and stroll away. Just *that* easy!

Upon arriving home after murdering Dawn Ashworth, Colin Pitchfork found a small drop of blood on his nylon jacket, two inches down from the left shoulder. He described how he cut a swatch the size of a match head out of the jacket. Then he realized that he'd lost his watch.

First he told Carole that he'd left the watch at work in his locker. The next day he told her it wasn't there, and the only thing he could figure was that he'd lost it while riding his motorbike. He did some worrying that the police would find that watch in the field and show a photo of it in the newspapers, but it never happened.

He did manage to bring home the food coloring on the afternoon of the murder. After strangling Dawn Ashworth, he baked a cake.

# 28

## Homage

. . . The psychopath shows a superficially adequate adjust-
ment. He is not anxious or distressed. . . . He shows no
blatant irrational thinking and displays no bizarre behaviors.
His initial charm and verbal ability distract attention from his
deviant and unfeeling behaviors.

—RIMM and SOMERVILL

Colin Pitchfork seemed in better spirits when it was
time to be interviewed by Derek Pearce and Gwynne
Chambers on other subjects. Even as a listener, Pearce
had an energy that generated conversation, and with
Chambers there was a sensitive tranquility that was
reassuring. The prisoner was so comfortable with these
two, he became more grandiose, and laced his speech
with macho profanity.

Colin Pitchfork told Pearce and Chambers that he'd
flashed a *thousand* girls in his lifetime. Ordinarily, he
talked in a monotone, but when he told of the flashings
he spoke with relish. Pearce decided to test him and
asked the prisoner to describe some that could be verified. Colin Pitchfork quickly ticked off three, and each
of them *did* check out. He bragged that he could spot a
good-looking bird three blocks away, and could cor-
rectly guess things about her at first sight.

He claimed to have flashed a girl in Cosby, figuring
her to be another hairdresser. He later followed her,
undetected, right to a beauty salon. He said newspaper

355

girls were easiest because they were so available. He described such triumphs with gusto.

Shortly after his confinement, Colin Pitchfork managed to secrete a shoelace, and he somehow removed a bolt from a brass plate in his cell. If it was an escape plot he never explained how he was going to escape with a shoelace and a metal bolt, but during the interview he produced them from his sock and put them on the table, expecting homage. The cops later said it must've been hard for him to pass through jail corridors wrapped in all that aura.

He described other crimes he'd committed, older crimes they weren't even aware were his: against the girl who'd walked home on a country lane, the girl he'd dragged back into the garage, the girl he'd picked up in his car and then dropped off after she'd grabbed the steering wheel.

He described many incidents, but in each there were little details he couldn't bring himself to admit, things like a screwdriver held to the neck of the young hairdresser. *She* told the police about that; he never did. He denied ever masturbating in front of the thousand girls he'd flashed in his lifetime and he was always quick to skip past any reference to premature ejaculation.

And he *never* changed his story that both Lynda Mann and Dawn Ashworth ran into danger on their own, through gates that opened on their own, while he was merely trying to let them pass. And he'd never verify the anal rape of Dawn Ashworth, which of course he may have done after he'd killed her. He'd never admit anything so unsavory, so *unmanly*.

Pearce always believed that Colin Pitchfork may well have had a weapon when he took both Lynda Mann and Dawn Ashworth into the fields, and Pearce never believed that either of them had engaged in small talk

during or after their rapes. A suggestion the prisoner
made that Dawn had actually joked afterward seemed
particularly grotesque.

Of course, rapists often claim that their victims didn't
resist them, even when physical evidence refutes such
claims, but Carole Pitchfork did verify that there could
not have been a prolonged struggle with either murdered girl.

She told Pearce, "You were always announcing you
were looking for someone with *wounds*. But he didn't
have a mark on him!"

Colin Pitchfork had an explanation for not trying to
rape and murder Liz, the blonde he'd picked up
hitchhiking.

He said, "I never touched her. Why? *That's* what
you're wondering, ain't it? The reason is, I was on me
own that weekend. Carole was always checking on me,
and if another one turned up dead, she'd start to wonder. I felt safe letting that one go because the blue Fiat
would look a different color to her under the sodium
streetlights."

But that explanation was totally at odds with a previous description of how he'd felt when he'd picked up
the blonde: *"Fuckin hell, Colin! This is your lucky
night, ain't it!"*

A more plausible explanation is that Colin Pitchfork
didn't attack Liz for the same reason that he never
attempted to murder the one person who controlled his
destiny: Ian Kelly. And for a sociopath, how insuffer-
able *that* must have been—to be under another's control.

Lynda Mann and Dawn Ashworth did what most
intelligent, even brave, people do when met with sudden overwhelming violent force: They reasoned, and
they believed they'd be spared. The girl he picked up
in his car did *not* reason. It wasn't a matter of greater

courage, just different instincts. Her grab at the steering wheel indicated to Colin Pitchfork that a death struggle was imminent. He could indeed get marked up by this one. He might lose control. It had already gotten dangerous.

Colin Pitchfork admitted that he'd *thought* many times about killing Ian Kelly, but said that if Kelly had ended up drowned, for instance, in a canal down by the bakery, there might be a link back to him.

Maybe, but there were lots of ways and other places to effect the murder of Ian Kelly if one had the stomach for taking on a full-grown man. Many sexual offenders, flashers in particular, have no taste for violence and pain, not when it's directed at *them*.

When he talked about his other crimes it was during an evening interview. He was more comfortable than ever with Pearce and Chambers, laughing and using more macho profanity: Everything was fuckin this and fuckin that. Halfway through the interview he decided he wanted a Chinese dinner. Derek Pearce and Gwynne Chambers had to cough up twenty-five quid for it. The prisoner *loved* that: coppers buying him dinner.

He'd spoken of the Scouts and of being in the same troop with a boy who later became a policeman in Coalville. He said he'd never flashed in Coalville for fear of seeing that copper.

When they asked about friends, he talked warmly of an old chum in Bournemouth as though they were still constant companions. But Mick Mason later located that man, rang him, and was told, "Colin Pitchfork? Yeah, I remember a bloke by that name. Haven't seen him since I was fifteen."

Mick Mason reported, "He had very little to cling to in the area of male friends."

Colin Pitchfork told them that he'd been far more nervous waiting for Ian Kelly during Ian's blooding on that cold night in January than he was during any of the interviews after his arrest. He was starting to sound like the kitchen porter: *Prison weren't too bad!*

As Colin Pitchfork was being brought to the interview room by Pearce and Chambers for one of their talks, a young constable standing in the hallway happened to step aside to let them pass. When their prisoner got inside the interview room, he said, "Did you see that? He knows who I am. Did you see the effect I have on people?"

The Monday after he was arrested, Colin Pitchfork was taken to Castle Court in Leicester for his first remanding. Castle Court looked exactly the way an old English court should look: stone walls and oak pillars, with arched windows going halfway toward a sixty-foot ceiling divided by blackened beams. Some of the ceiling beams were original, among the oldest in Europe, dating from A.D. 1105 when the building had been an armory. A cracked and buckling oil painting hung high over the bench, and the county magistrate sat beneath a carved canopy at a most commanding height. It seemed a proper place for a remanding on this, the most massive police inquiry in the county's history.

The stone entrance to the castle yard was part of a fortified gate that used to surround the armory, and portions of it dated from the 14th century. The drive to the yard was over cobbles, past brass streetlamps. Waiting for him there were reporters and television crews, as well as a small crowd, mostly women, who booed and jeered and shouted threats and oaths as the CID cars passed, escorted front and back by marked police cars. Colin Pitchfork, unkempt and unshaven, bent forward, his head hidden under a blanket while they screamed things like "Cowardly bastard!" and "Bring back hanging!"

The twenty-seven-year-old defendant, who, at his best, could project a sardonic air, seemed disdainfully attired in jeans and a casual shirt. He was represented at the remanding by Walter Berry, Tony Painter's acquaintance, who seemed to represent everyone connected with the case. The defendant was charged with the two murders as well as two indecent assaults: on the girl he'd pulled from the country lane in 1979 and the one he'd dragged into a garage in 1985. He was also charged with kidnapping in the case of Liz, the girl who'd grabbed his steering wheel.

He was in the dock only long enough to answer that he understood the charges, then he was whisked back to the lockup. As he was driven out of the castle yard, people in the crowd yelled, "You bloody murderer!"

Hiding under the blanket, handcuffed to Mick Mason, the prisoner said, "Yeah, *that's* right!"

---

There was a shock awaiting Derek Pearce on the 23rd of October. He'd completed the court file on Colin Pitchfork at 12:10 P.M. and submitted it for final vetting, thus officially ending the inquiry into the footpath murders. At 2:10 P.M. he was *suspended* without pay from the police force.

Despite earlier assurances to the contrary, he was going to receive a summons to criminal court and be prosecuted for causing actual bodily harm to the policewoman.

When Pearce left the office of the deputy chief constable after turning in his warrant card and key, Chief Supt. David Baker brought him into his private office.

Baker poured Pearce a huge Scotch and they had a long chat, but Pearce was too upset to do more than catch the drift of it. Baker said he'd been unaware they were going to prosecute Pearce. Now he was obviously afraid for his DI. From the look in Pearce's eyes, Baker just couldn't be assured that Pearce wasn't going to do something crazy.

"Tell you what I'm going to do, boss," Pearce said finally. "I'm going to collect my things and bugger *off*."

"Sit down and talk some more," Baker urged.

"I've got no ax to grind," Pearce said. "I'm *going*."

"*Please* sit down a bit," Baker said.

But Pearce thanked Baker for the drink and walked out.

David Baker immediately ordered Mick Thomas to babysit. He said to Thomas, "You're to stay with Derek until he goes to bed. You're to *put* him in bed, if necessary!"

But this was one baby who wouldn't behave. Pearce ran to his black Ford, popped it in gear, and squealed out of the police car park before Mick Thomas could catch up. Baker ordered Mick Thomas to find him, but Pearce spent the evening safely in a pub with Gwynne Chambers and Phil Beeken who collected and delivered his remains at evening's end.

The internal investigation against Derek Pearce quickly turned nasty. It seemed to Pearce the investigators were trying to prove that he'd been a discredit to the police force since 1983. That was a bit hard to do since he'd been helping to ramrod the Narborough Murder Enquiry since then. Nevertheless, they went back in his police diary and prepared a discipline form loaded with infractions such as "ripping up police forms with the intent to destroy." Things like that.

While it was being proved that Pearce was a discredit
to the force, and while he required a uniformed escort
even to enter a police station, they discovered that they
needed him again. Mick Thomas probably wasn't as
good on paper as Pearce was, and there were problems
with the Pitchfork court file.

Mick Thomas was sent to Pearce's home to learn how
statements of Colin Pitchfork fit into sequence. Pearce
was too much of a cop to refuse.

But they wouldn't let Derek Pearce's solicitor interview any police witnesses without a senior police officer
being present. In short, he was denied the advance
disclosure that an ordinary criminal is granted. And a
colleague who'd witnessed a *prior* physical row between Pearce and his policewoman accuser in a pub
declined to testify on his behalf.

Pearce obviously felt surrounded by disloyalty and
betrayal. Cambridge in the '30's hadn't spawned so
many traitors.

For those in the Leicestershire police who may have
been eager to extinguish this Roman candle of a cop,
Derek Pearce made the job a bit easier. After attending
a rugby match on December 26th, he offered to drive
home a friend who'd been imbibing too much. During
the ride, according to Pearce, the friend made a rude
gesture to another driver, and while Pearce attempted
to quiet him, he swerved. The angry motorist took
down Pearce's registration number, and four months
later he discovered *another* criminal charge added to
his first: driving without reasonable consideration for
other road users. This meant that Pearce would be in
Crown Court *and* Magistrate's Court on the same day
in June, 1988. Not many had accomplished such a feat.

The embattled detective got himself a good barrister,

a man known as a tough advocate in criminal cases. Pearce said it was impossible to imagine himself on the *other* side at a criminal trial.

It was unusual to find a red-robed judge adjudicating minor infractions like those Pearce faced, but Pearce was a police inspector, a controversial police inspector. His prosecution was being vigorously pursued.

Pearce had repeatedly told friends that he was innocent and couldn't even *conceive* of a conviction and imprisonment. But when he learned what sort of judge would be assigned to adjudicate his case, he knew that he and Colin Pitchfork had something in common. They'd both be facing a red-rober.

# 29
# Outrage

When confronted with his misconduct the psychopath has enough false sincerity and apparent remorse that he renews hope and trust among his accusers. However, after several repetitions, his convincing show is finally recognized for what it is—a show.

Nearly every type of treatment method has been tried with the psychopath. In general, the treatment . . . has not been rewarding nor enlightening.

—SUINN

The first snow of winter fell on January 22, 1988, as Colin Pitchfork was driven to Crown Court in Leicester inside a van with blacked-out windows. The prisoner, who'd grown a full beard, was rushed from the van into the courthouse, blinking his eyes in the watery winter light.

That courthouse isn't old and steeped in history like Castle Court. It's red-brick modern with tinted windows, so serviceable and boring that graffiti might improve it. But the courtrooms are large enough to accommodate about a hundred people. The spectators, mostly press, all queued to pack themselves inside.

Except for a coat of arms behind the bench, the courtroom was stark, but there was a red-robed judge, black-robed advocates, and bleached horsehair wigs to add a note of dignity to the Crown Court's Holiday Inn decor of white oak and earth tones.

The national press wanted the family reaction.

Barbara Ashworth said, "I *had* to come. I had to see him. To lay the ghost to rest."

Eddie Eastwood spoke for himself and Kath, saying, "We had to go. I just wanted to see his face. I wanted to know what sort of man could do it."

It was decidedly anticlimactic, that sentencing of the Narborough murderer. The judge had several serious cases to deal with and didn't appear to place special emphasis on this one.

Ian Kelly, referred to by the police as an "extremely gullible person," was told by the judge, "I just about believe you did it because you accepted the story put forward by Pitchfork."

Ian was given an eighteen-month prison sentence, suspended for two years, which meant he would not have to serve time. When Ian stepped from the courtroom, trembling like a whippet, he wiped his eye. And with his sturdy young wife on his arm for support, he said to the television cameras, "I was wrong for doing what I did. I'm sorry for whoever I've caused grievance to. And I'm, well, *shocked!*"

The prosecuting barrister, Brian Escott-Cox, Q.C., read a summary of Colin Pitchfork's crimes from the court file. He added that the defendant "showed amazing self-control with a total lack of remorse" in that not even his wife had had any idea he was a killer.

Colin Pitchfork was dressed in summer clothes: jeans and a short-sleeved shirt. He pleaded guilty to the murders of Lynda Mann and Dawn Ashworth, and to those two indecent assaults he'd revealed from out of his past, and to conspiracy to pervert the course of justice by his use of Ian Kelly. He pleaded not guilty to kidnap in the case of the hitchhiking girl whom he'd "spared."

By way of mitigation, David Farrer, Q.C., the defendant's barrister, said, "He recognizes he can do nothing to alleviate the overwhelming suffering and grief inflicted on the families of his victims by his frightful evil—and those are his words, not mine."

The barrister then added, "He will remain forever haunted by the images and knowledge of what he has done."

The court's psychiatrist no doubt could've explained that Colin Pitchfork's concept of being "haunted by images" was very different from that of his barrister. When the defendant had talked to Mick Mason and Mick Thomas of being "haunted" by Lynda Mann, it was with bemused detachment. The "hauntings" of a psychosexual sociopath provide not horror but *inspiration*. Probably his barrister could not envision the ecstasy such hauntings would bring.

Colin Pitchfork received a double life sentence for the murders, a ten-year sentence for each of the rapes, and three years each for the sexual assaults in 1979 and 1985, along with another three years for the conspiracy involving Ian Kelly. These were concurrent sentences, and much to the astonishment of the murder squad, the judge *didn't* give a recommendation for a minimum term. Without such a recommendation, the "life" sentence in Britain was similar to that in the United States, which meant that Colin Pitchfork could conceivably be released in ten or twelve years. The police were outraged.

While passing sentence, the high court judge, Mr. Justice Otton, said, "The rapes and murders were of a particularly sadistic kind. And if it wasn't for DNA you might still be at large today and other women would be in danger."

The judge added that a psychiatrist's report compiled by a Broadmoor doctor diagnosed Colin Pitchfork as a "psychopath of a psychosexual type." And the judge said that the defendant would receive therapy in prison for his condition and would not be released until that therapy was complete, which would be in "many years."

That led a few observers to note that if prison doctors had found effective therapy for a sociopathic serial killer, they should patent it and eclipse Alec Jeffreys's discovery of DNA fingerprinting. A "cure" implies change, a discovery of contrition. But to a sociopath the absence of a crippling emotion like remorse is a blessing, not a curse.

Eddie Eastwood later said, "Pitchfork looked at me, eye to eye. He just stared me out as if to say, 'Well what's the matter with *you*?' I couldn't make him out. He looked almost human."

Kath said, "It was the shock of seeing him. The shock! I didn't look up when the lawyer passed those photos of Lynda to the bench. Those photos of how she looked when they found her. The cover dropped open and the audience gasped when they saw the photos. My brother saw them and cried. Luckily, I didn't look up."

The mother of Dawn Ashworth wanted a trial, a real trial, with trappings and *finality*. Colin Pitchfork didn't have a real trial, she later said, just a hearing. With a trial he'd be exposed for what he was, she thought. A guilty plea seemed just a clever ploy to avoid real exposure. She listened in amazement as the prosecutor read the summary of his crimes, how he'd killed Lynda while his baby lay in a carrycot in the back of his car.

And the most horrifying moment was his description of Dawn sitting up, having a conversation, almost *joking* with him, after all the things he'd done to her. Only to die piteously when she thought she was going to be spared. The Ashworths were utterly devastated by that testimony, and grateful to learn that detectives usually hear such self-serving stories from rapists. They were thankful when the judge read from the pathology report that Dawn had been close to death when the killer viciously violated her.

Carole Pitchfork, sitting next to a policewoman, leaned forward in court to get a look at Barbara Ashworth who was accompanied by Robin and Supt. Tony Painter.

Barbara had a feeling that the plump young woman must be Colin Pitchfork's wife. She asked Tony Painter and he verified it.

Barbara's emotions were rampant. She couldn't believe that Colin Pitchfork hadn't been marked during Dawn's murder. Dawn had been so proud of her nails. Carole *must* have known, must have at least suspected, must have shut her eyes. Or maybe *not*. Barbara just didn't *know*. Not knowing could be the cruelest, sometimes. Except for the outrage. To survive one's murdered child. The infinite outrage.

To Barbara Ashworth, Dawn's wristwatch was sacred. She'd never gotten Dawn's clothes back because they were used in forensic tests, but she did get the watch. Barbara had worn it to Australia, always keeping it set to English time, and had worn another with Australian time.

Barbara said of the watch, "It had *never* stopped and it kept perfect time, and maybe it sounds macabre, perhaps it's sick, but that watch lay there *with* her for those two days. . . ."

So when the mother of Dawn Ashworth sat in that courtroom and looked at *him*, she touched the watch frequently.

It was staring at him and having him look straight *back* at her, without contrition, that made her suddenly cry. Like all the others, she was searching for something in him but did not find it. He looked nothing like the spiky-haired punker she saw in her nightmares. His receding hair and beard were the color of farmhouse eggs. His face was chubby and expressionless.

Standing there in the dock he could occasionally look sardonic, yes. But he was really so ordinary. So *banal*.

Colin Pitchfork's sentencing had been slotted in on a busy day. He went in at 11:50 A.M. and they adjourned for lunch at 1:00 P.M. They reconvened at 2:15 and continued until 3:00. And then it was over. Yes, it was decidedly anticlimactic.

And it was a pity that the psychiatrist didn't choose to describe him as a "sociopath" instead of a "psychopath" in his report, because of the misunderstanding that accompanies the latter. Everyone connected with the case seemed to confuse the word with "psychotic." Even the journalists made the mistake, writing copy like "Only when brave Liz grabbed hold of the car's steering wheel in an attempt to force it off the road did Pitchfork snap out of his *psychopathic* trance and agree to take her home."

Both the television and print media showed pictures of Colin Pitchfork at his wedding, sporting a silk topper, his brows arched with a saw-toothed grin, all hinting of a latter-day Mr. Hyde. There were many references to his "sickly grin" or "dead eyes," and endless allusions to "evil." Almost everyone, it seemed, preferred original sin to clinical definition.

And everybody was distressed by his indifference, though it was totally consistent with the tendency of sociopaths not to respond to threatening events as normal people do. The physical indicators of stress and apprehension just weren't there, which explains why sociopaths are unfit subjects for polygraphs.

So while the defense lawyer spoke of a haunting, and the judge talked of treatment, the fact remained that

Colin Pitchfork may have understood instinctively that he could no more alter his makeup than he could alter his genetic fingerprint. In his interview with Derek Pearce and Gwynne Chambers he told them he hoped to study accounting while in prison. He wasn't dismayed by a prison term. He said, "I'll simply be changing a larger world for a smaller one."

The judge and the defendant's barrister implied that a third act could be written. But for the sociopath there is no third act.

# 30
# Hindsight

Lykken (1957) demonstrated that psychopaths do not develop the fear necessary to avoid a noxious stimulus. . . . They simply do not learn well from punishment, an observation that indicates imprisonment will not change their behaviors and personal traits.

—RIMM and SOMERVILL

. . . Little is really known about possible organic factors that might be involved in the psychopath's impulsive behavior. Until the etiological picture is clarified, systematic therapeutic procedures will be difficult to develop.

—SUINN

When it was over, the British media offered lots of the tabloid quotes for which there are no equivalents in the rest of the world. Such as: "Behind his sickly smile was the evil mind of a killer," or "His deviant mind was to plummet to new depths."

Carole Pitchfork was bitter about the press coverage. One story claimed that the life of the hitchhiking girl was spared because Colin Pitchfork had suddenly realized supper was ready and he had to rush home to a wife who was a strict disciplinarian. This, even though the crime had occurred at one o'clock in the morning when Carole was on a camp-out with the kids.

Eddie Eastwood was interviewed by the television news and claimed that after seeing Pitchfork and Kelly,

377

he realized he'd played darts on Christmas Eve in a Whetstone pub with the two of them. But police found this very unlikely. Both Colin Pitchfork and Ian Kelly said they'd never been together outside of work, except during the blooding scheme.

As to his feelings about Colin Pitchfork, Eddie said, "I'd like him to be in front of me, so I could bleed him dry very slowly. Hanging is the *only* way to deal with this monster."

Kath said, "He must never be allowed to walk the streets again. He should *hang*. With this new DNA genetic fingerprint there is no chance of a person being later proved innocent after he's been hanged. There is no excuse anymore."

The television reporters wanted an interview with the Ashworths but the Ashworths declined. However, an old interview, given long before Colin Pitchfork had been captured, was intercut with footage dealing with his sentencing.

The old segment showed the Ashworths when they were still working with the police, attempting to pique the conscience of the killer's family. In that old interview the reporter asked, "Is it conceivable that you might forgive the man, in your own heart?"

There was a very long pause on that videotape before Barbara Ashworth could swallow and say, "I *have* to. Because otherwise you'll spend the rest of your life being very bitter and twisted. And I don't think you can go on like that."

During that old interview Robin said, "I don't feel any hate or wish for any revenge for the murder of Dawn, because it's not going to do any good, whatever I feel. It's not going to do *any* good."

In February, 1988, when the new show was aired, the announcer told his viewing audience, "The parents of Dawn Ashworth have no bitterness toward the man who robbed them of part of their lives, only forgiveness."

So the Ashworths saw themselves on TV offering absolution to Colin Pitchfork. They'd suffered every other indignity, now humiliation.

As to how they truly felt, Robin said, "If the genetic test can prove guilt beyond the shadow of a doubt, I don't see why they don't reintroduce the death penalty."

One month after Colin Pitchfork was sentenced, the *Leicester Mercury* polled its readers on a proposed return to capital punishment, and one reader in ten responded. The respondents felt a need for the Lord of Death with icy breath. Ninety-six percent wanted to bring back the hangman.

There was a fair amount of hindsight and second-guessing to be found in news reports, and Chief Supt. David Baker faced a grilling about alleged "blunders." Journalists wanted to know why, given Colin Pitchfork's flashing background, he'd never been brought in for serious interrogation, and how an altered passport could slip past the police. Baker explained that Colin Pitchfork hadn't even lived in the village when he'd killed Lynda Mann, and there'd been thousands of people giving blood and presenting all sorts of identification to harried detectives. He ended by saying he could offer no guarantees when dealing with deceptive criminals.

There had been a critical editorial asking why Colin Pitchfork had never been photographed and fingerprinted during his earlier brushes with the law when he'd been charged with flashing offenses. Baker said that in years past the police had not routinely photographed and fingerprinted minor offenders. The law had always treated flashing as a nuisance crime.

At the end of the day, it had to be said that nothing

would have changed even if there *had* been photos and
fingerprints taken in Colin Pitchfork's early flashing
career. Latent prints had not been found at the scene of
the Lynda Mann and Dawn Ashworth murders. And it
was unlikely that a prior photograph would've been
pulled from his file to await his arrival at the blooding.

Moreover, the girls he'd assaulted in 1979 and 1985
and 1987 probably would never have picked out his
mug shot among those of hundreds of Leicestershire
sex offenders, since in one episode he had grown a full
beard and the other two happened in the dead of night.
In any case, an arrest for an earlier assault would not
necessarily have diverted Colin Pitchfork from violence.
He had always been more opportunistic than compul-
sive, this sexual sociopath.

Chief Supt. David Baker said that he was satisfied
with the way his men had conducted both murder
inquiries with untiring self-sacrifice. Despite the clamor
for a return to hanging, David Baker still wasn't sure
about capital punishment. He worried as to whether
some killers truly had the capacity for criminal intent as
defined by law.

As for the other senior officers, Chief Supt. Ian Coutts,
the Scotsman who'd commanded the Lynda Mann inquiry, said one evening at police officers' mess, "I'm
not exceptionally religious, but I believe God had a
hand in this DNA business." Supt. Tony Painter, commander of the Dawn Ashworth inquiry, completed thirty
years of service in February, 1988, and retired.

The bakery outlet manager came to the police station
amid fanfare. The police brass had decided she would
be awarded half of the £20,000 reward for having reported Ian Kelly's pub chat to the police. The other
half was not awarded.

There had never been an investigation like it, and

the future of genetic manhunts was now being studied by lawmen from all over the world. The revolutionary murder hunt for the footpath killer had resulted in the blooding of 4,583 young men, the last being Colin Pitchfork of Littlethorpe, whose DNA pattern did indeed provide a perfect match to the genetic signature left by the slayer of Lynda Mann and Dawn Ashworth.

---

The older son of Colin Pitchfork missed his father. Carole explained his absence by saying, "Daddy had to go away. He can't come back but he loves you a lot."

When the child asked why his daddy had to go away, she said, "For doing something bad."

"Like breaking a window, or something?" his son asked.

"No," she explained, "something really bad. Like hurting somebody, or something."

When the child painted pictures he'd often say, "This one's for Daddy."

Carole often wrote letters *for* the boy and sent the paintings to her husband, but she never wrote to him on her own. Colin Pitchfork in turn composed picture stories for the children, stories that Carole would not share with any adult. He drew maps and animals indigenous to various countries, and he'd tell a brief story about that country, such as one about Eskimos and a whale as big as a double-decker bus. At the end there'd be questions: How big is a whale? What does a whale eat? Carole would read the stories aloud.

Carole's failure to visit or write to her husband caused trouble with her mother-in-law who said, "You *ought* to go see him."

"If he was my son, I would," she told his mother. "I respect you and you *must* respect me." But finally

Carole lost her temper, after which they had a very strained relationship.

Carole had become a part-time youth worker for the county council. It was financially and emotionally draining, raising her young children all alone. And sometimes there was *too* much time to think.

"It's like somebody being dead," Carole said. "Or perhaps like being at a seance, because letters come in the letter box, as if from another world. It'd be better if he *was* dead, then you could grieve and get over it and it'd be finished.

"His parents want to see him free before they die. But I feel dread at the thought that someday, years from now, he'll knock. Yet other times I forget he's gone. Whenever I'm at home and hear a motorbike, I expect him to come in the door."

------

After all the smoke had cleared, the last sixteen members of the murder inquiry had a party at a local pub. Derek Pearce presented the others with personal gifts that related to intimate moments during the long inquiry. One of them had gotten a bit tipsy during a mid-inquiry do and had tried to scoop a fish from the aquarium of a seafood restaurant. Pearce presented him with a fishbowl and a live goldfish to commemorate the event. Another, who'd suffered an eye injury during one of their more raucous parties, was given the fresh eye of a bullock.

Gwynne Chambers's old passport was redone with a picture of the back of John Dayman's head and the signature "Colin Pitchfork." It was presented to the cop who'd been fooled by Ian Kelly.

They knew there would never be another case so

important in their careers. Many of them reported a letdown, of feeling bored and unsettled after returning to ordinary police work. The intensity was missed. Some of them became quieter and more serious.

The goldfish ended up with John Dayman, who kept it as a pet and named it Colin.

By mid-1988, United States lawmen had gone mad over genetic fingerprinting, and variations called DNA typing or DNA fingerprinting. There were criminal cases involving genetic fingerprinting prosecuted in Florida, Oklahoma, New York, Pennsylvania, Virginia and Washington—all with positive results—and large companies in Maryland, California and New York were doing genetic fingerprinting analysis. There were frequent network television reports as well as newspaper and national magazine stories about one criminal case or another having been solved by the new forensic miracle.

The American Civil Liberties Union was studying whether computer data banks, or even voluntary DNA testing, raised constitutional issues, as well as whether DNA information could be used to persecute AIDS carriers, or even for DNA fishing expeditions.

Despite the contentiousness of the American people and their elephantine legal system, one thing seemed certain: The technology discovered by Dr. Alec Jeffreys was able to withstand the scrutiny of molecular biology scientists. Genetic fingerprinting was here to stay. A new industry had been born. Its future and possibilities seemed unlimited. Experts predicted that someday an American citizen's bloodprint would be as accessible as his Social Security number.

During his eight-month suspension while he awaited trial, Pearce didn't spend much time cooking for police friends, because suspended officers were forbidden to associate with other policemen. He spent more time socializing at his favorite pub which was seventeen miles from home. A pub with no other police patrons.

He had a lady friend whom his police mates had never seen. There were rumors that she'd suffered some sort of serious injury, and that Pearce had taken care of her, even going so far as to urge her to accept private medical treatment at his expense. None of his former colleagues on the murder squad expected to meet her. If *this* one ended up flying east they'd never know about it. Humiliation would not be added to heartache.

Now that he was stripped of his rank and authority, Pearce could no longer hide his insecurity behind an aggressive, abrasive façade. So he substituted secrecy, and had a genius for keeping the world at bay. Always a telephone hater, he became harder to reach than Marlon Brando. No one was quite sure what he did to keep the larder stocked and a roof over his head, but he seemed to manage and to stay busy at unspecified tasks for unspecified associates. Perhaps he felt more of a pariah than he would ever admit, especially to himself.

Just as he couldn't be a good patient for a doctor, he couldn't be a good client for a lawyer. He'd never consider mitigating circumstances. On the night of the alleged assault on the policewoman, he'd just come from the meeting at Tony Painter's house at which the entire inquiry faced the prospect of shutdown. Everyone *else* on the murder squad admitted feeling anger, frustration, resentment. In a word: stress.

Pearce couldn't accept that. He insisted, "I never suffer from police stress. I *need* it!"

His idea of admitting personal weakness was to say, "They call me arrogant because I don't suffer fools gladly."

Pearce's barrister and his solicitor probably realized that they might have to go to trial against a most hostile witness, one capable of doing terrible damage to their client. The witness named Derek Pearce.

During the course of the trial, Derek Pearce's solicitor cautioned him about his apparent contempt for the proceedings. Even the judge remarked that the defendant showed a "cavalier attitude." Nevertheless, on July 5, 1988, after a five-day criminal trial, a jury retired for only forty-eight minutes before returning a verdict of not guilty on all counts.

Derek Pearce was reinstated to the police force and all internal disciplinary proceedings were dropped. The deputy chief constable said that the matter of back pay was "negotiable," and Pearce said that charges should never have been brought. He was to take a short leave and then be transferred to uniform duties "to be kept out of the public eye."

Chief Supt. David Baker's secretary told Pearce that when Baker got word of the acquittal he danced around his desk and ran off to find and congratulate his willful young inspector. Pearce vowed he'd be back in CID working for David Baker as soon as he'd done sufficient "penance."

Of course, Derek Pearce would never admit to having been frightened or worried about the criminal charges he'd faced, not he who professed to *thrive* on stress. But during all those months when he'd been stripped of rank and authority and forbidden to associate with other police officers, there had been moments when his defenses wavered. Once or twice he'd even admitted to some vulnerability.

Besides police work, what he'd painfully missed was police officers' mess on the first Monday of the month, wherein senior officers meet in formal dress for drinks and dinner. He'd bought himself a tailor-made dinner suit, and he loved the jokes and speeches and camaraderie, the sense of belonging to a band of brothers.

"It's ever so grand," he said, "to raise a glass of port and toast the queen."

# 31

# Churchyards

At the Narborough end of The Black Pad is an ancient church and cemetery. All Saints is High Church, with a mossy slate roof and granite turrets that glow rose and amber under curdled skies. On Sunday evenings church bells announce evensong, and in the summer, stained-glass panels emit ruby and emerald rays through a Gothic arch when the light nights bring bronze and copper sunsets to the Midlands. The old headstones are furry with moss and the footpath bricks laid on end have been worn smooth by the feet of villagers who have sought continuity there over the centuries.

Kath had never been one to visit graveyards. "I'm not much of a believer in grounds," she said. "Still, it's an outing, to visit the grave."

Kath and Eddie Eastwood found they didn't like living away from Leicestershire after all. When Lynda's killer was finally captured they decided to return. They were still afraid to let their daughter, Rebecca, walk alone in the lanes, but at least the village felt like home.

Old reliable Mick Mason showed up at their door with a bottle of sherry to welcome them back.

By the motorway, on the east side of Enderby near Ten Pound Lane, is another churchyard and another age-blackened granite church that seems as enduring as time. There, brown sparrows and gray wheel on the wind only to plummet and perch on headstones. Black leaves scratch at the graves and the older headstones are covered with lichen, nature's reminder that flesh is grass.

Inside that church the smell of hymnbooks, the sound of holy music, the casting of Rembrandt shadows in Midlands silver light all promise continuity. Though Robin and Barbara Ashworth were only occasional parishioners, the canon of St. John Baptist offered them a cemetery plot next to Dawn, and they took it gratefully.

Robin and Barbara reported that they'd begun changing in subtle ways. Robin said he didn't need as much sleep anymore, and was no longer tense and nervous at work. He just didn't let little things bother him. Robin had learned to sort out what was worth the worry.

Barbara said she'd learned to put herself on automatic pilot to get through some days. Dawn's room remained just as she'd left it, but sometimes Barbara borrowed Dawn's clothes.

"Once I needed a jumper to go with gray," she said, "and Andrew told me to wear that one of Dawn's. But because it was one she wore a great deal, I wouldn't have wanted to wear it. I remember once when I wore something my mother handed down to me, Robin came in the kitchen and said, 'Oh, I thought that was your mother standing there!' I didn't want to wear something of Dawn's and give him the same feeling. So I won't wear that jumper."

Then she said, "But I secretly wear it sometimes."

## About the Author

JOSEPH WAMBAUGH, formerly of the Los Angeles Police Department, is the author of fifteen previous books—*The New Centurions, The Blue Knight, The Onion Field, The Choirboys, The Black Marble, The Glitter Dome, The Delta Star, Lines and Shadows, The Secrets of Harry Bright, Echoes in the Darkness, The Blooding, The Golden Orange, Fugitive Nights, Finnegan's Week,* and *Floaters*—all of them outstanding bestsellers. He lives in southern California.

## THE WORLD OF
## Joseph Wambaugh

•

*His fiction reads like truth, and his True Crime is as grip-*
*ping as the most compelling novel. He is Joseph*
*Wambaugh—and no one knows the dark, gritty side of*
*cops like he does!*

•

# THE GLITTER DOME

•

*Welcome to Tinsel Town, where the line between cop and*
*killer is a razor's edge ... Movie mogul Nigel St. Claire*
*has been murdered and for the cops of The Glitter Dome,*
*a teeming, loud smoky watering hole, their seduction by*
*Hollywood is about to begin. Yet there is more than money*
*and glitz in this world as veteran cops Al Mackey and*
*Martin Welborn will discover.*

At first, Herman St. Claire III stared at Al Mackey
blankly when he held out his hand. "Mackey and Welborn,
L.A.P.D.? Remember?"

"Oh sure!" he cried. "Sure. Al and . . ."

"Marty."

"Of course! So glad you could come! I'd like you to
meet . . ."

But the Famous Singer had boogied as soon as she
heard who they were. They weren't like the cops at home
in Queens. These L.A. cops would bust their mother if she
snorted one spoon. And the Famous Singer had a Bull
Durham tobacco bag around her neck under her sweat
shirt clearly stenciled: "Nose Candy!" The bag was full of
cocaine that cost $150 a gram and was guaranteed to be
quality stuff that wouldn't embarrass her at a nice party.

Everyone who saw it said it was a darling idea too. No way was she going to let some cop confiscate it.

"Listen, I'll introduce you around if there's anyone you wanna meet. Meanwhile you boys help yourself and mingle." Then, as an afterthought, Herman III said, "Oh, by the way, you getting anywhere on my uncle's case?"

"Not much happening yet," Al Mackey said.

"No? Too bad. Listen, you fellas mingle."

Al Mackey saw a bizarre art deco costume of graphic zigzag, red line on white, done in folds and wraps and ending up with a puffy mini over leggings. It was topped off by a hat-helmet with simulated strands of gold brocade hair. And then he recognized the girl: Tiffany Charles!

Martin Welborn began nibbling at one of the ordinary items on the mile-long table, baby shrimp in guacamole sauce, when he turned to see Al Mackey trotting across the dance floor, his second tumbler of whiskey giving him the courage to burrow right through a crowd and say, "You're Tiffany, Mister St. Claire's secretary. It's *me*, Al Mackey. Sergeant Al Mackey? Remember?"

"Oh yeah," she said. "I really don't know anything more tonight than I did the other . . ."

"This is a *social* occasion!" Al Mackey cried. "I *love* your outfit. I've never seen gold hair. Is it real?"

"Uh huh," she said, seeking rescue. Already her friends were drifting away.

"Fourteen karat?"

"Twenty-four," she muttered.

"Wow! They pay secretaries pretty well where you work."

Somebody save her from the scrawny cop! A dress like this he thinks you earn taking dictation? Help!

"Listen, I gotta go talk to some of Mister St. Claire's stars," she said. "You just *mingle*, huh? Have a good time." "I'm *trying* to mingle," he cried.

# THE DELTA STAR

*In Wambaugh's world of L.A. cops, the suicide of a cheap
hooker from the roof of a sleazy hotel seems like a normal
desperate leap from life. Yet nothing is ever as it seems,
thinks Chip Muirfield, just one of the cops of Rampart Station who, along with Mario Villalobos, The Bad Czech,
and Jane Wayne, follow the trail of corruption from the
world of pimps to the country's top-ranking chemistry
wizards—and beyond.*

This was only the third autopsy that Chip Muirfield had
ever witnessed. He enjoyed each one more than the last.
Mario Villalobos thought that if Chip started liking them
any better, the kid might start moonlighting at Forest
Lawn. The pathologist and technician were trying like hell
to get this one zipped in time to watch *Days of Our Lives*.

The former Western Avenue prostitute, who had delighted
Chip Muirfield by dying not in Hollywood Division where
she worked but in Rampart Division where she lived, was
not broken up too badly by the fall from the roof, at least
not her face. Mario Villalobos thought of the early mug shot
of this face now peeled inside-out like a grapefruit. A natural blonde, fair and slight; he wondered if she drove them
wild when she got that tattoo of the man-in-the-moon. It
was on the inside of her left thigh, high enough to have
been a very painful job. In death she looked thirty-five years
old. Her identification showed her to be twenty-two.

Mario Villalobos was one of those homicide dicks who
somehow revert to uncoplike sentimentality during mid-
life crisis. That is, Mario Villalobos, like his old partner
Maxie Steiner, gradually came to resent needless mutila-
tion of corpses by cutlass kids who, quite naturally, are ex-

tremely unsentimental about carcasses in which detectives
have a proprietary interest.

What Mario Villalobos didn't see while he was roaming
the autopsy room, thinking of how dangerous it is to go to
The House of Misery every single night, was Chip
Muirfield's interest in the man-in-the-moon tattoo high up
on Missy Moonbeam's torn and fractured femur, close to
the inn-of-happiness which the bored pathologist figured
was *really* what was interesting the morbid young cop.

It was a professional tattoo. The man-in-the-moon had
winked one eye at Chip Muirfield and with the other
glanced up at the blond pubis of Missy Moonbeam. It was
a very cute idea, Chip Muirfield thought, but the leg was
so destroyed by the fall that the upper thigh was ripped
open and hanging loose.

"I wish it weren't so damaged around that tattoo. It's all
ragged and bloody and it's hard to see. Snip it off there
and I'll have the photographer come and shoot a close-up
of it that we can use."

Mario Villalobos returned and noticed that Chip
Muirfield was so intensely interested he looked ready to
crawl inside Missy Moonbeam.

Mario Villalobos looked at the butter-brickle three-piece
suit worn by Chip Muirfield, hesitated a moment, and then
said, "Even Boris Karloff wasn't so eager, Chip. If I were
you I'd step back just a bit."

But Chip Muirfield didn't seem to hear him, so Mario
Villalobos went for coffee. The pathologist pulled off his
gloves and called it a wrap. The technician looked up at
the clock and . . . Jesus Christ! *Days of Our Lives* was going to start in three minutes!

That did it. He reached for the faucet over the gut pan
to get this baby zipped. He wasn't paying any attention to
a young surfer-cop in a butter-brickle suit. He was eyeing

that clock like a death-row convict and he cranked the faucet full blast. The water hit the gut pan with a crash. And Chip Muirfield was *wearing* Missy Moonbeam.

His butter-brickle three-piece suit was decorated by a geyser of blood. A piece of Missy Moonbeam was plastered to his necktie. Another little slice of her hit him on the lapel. A swatch of Missy Moonbeam's purple gut plopped on his shoulder and oozed like a snail. But worst of all for Chip, who was yelling and cursing the technician—who couldn't care less—Chip Muirfield had a wormy string of Missy Moonbeam's intestine dangling from his sunburned surfer's nose.

•

# THE SECRETS OF HARRY BRIGHT

•

*Seventeen months ago Jack Watson was found incinerated in a Rolls-Royce, a bullet in his head. Now, L.A.P.D. homicide detective Sidney Blackpool is called into this still unsolved case and the investigation soon becomes an obsession, as memories of his own son's death plague him.*

Sidney Blackpool chain-smoked all the way back to the hotel. Otto had to open the window to breathe, shivering in the night air that blew through the canyons.

"Making any sense yet, Sidney?" Otto finally asked.

"I dunno. Sometimes part of it does, then it doesn't."

"A dope dispute? Naw, we ain't talking big dopes. How about a straight rip-off by the Cobras? They set up the gay boys in the bar, bringing them up the canyon with a promise of low-priced crank and waylay them."

"Why two cars then? Why was Jack Watson in the Rolls and Terry and the marine in the Porsche?"

"Yeah, and why wouldn't Terry step up and tell his

story right away if he saw somebody kill his pal? Especially after the reward was posted."

"Maybe he was already outta town by then. Anyway, Billy Hightower says he's sure his people didn't do it. Billy *does* seem to have effective interrogation methods."

"And what's Harry Bright got to do with it? And why's Coy Brickman nosing around out there now that we're stirring things up?"

"There's always the possibility that Terry planned the kidnap and ransom of his pal Jack with the help of Bright or Brickman," Sidney Blackpool said.

"Should call this place *Urinal Springs*, you ask me," Otto said. "The whole place stinks, far as I'm concerned. It's like the city of Gorki, closed to foreigners. And we're foreigners, baby."

"First thing tomorrow let's work on the uke. We'll call the manufacturer. See what they can tell us. I wonder how many music stores there are in this valley? Not many, probably."

"It's hard to imagine Harry Bright involved in a murder, ain't it?" Otto said.

"You never even *met* Harry Bright."

"You're right. This place is making me goofy. It ain't real hard to think a Coy Brickman icing somebody down. Those eyes a his, probably the freaking buzzards got eyes like he's got."

"We gotta get this connection between Harry Bright and Coy Brickman. Maybe it started back in San Diego P.D."

"What?"

"Whatever might make one a them or both a them kill Jack Watson."

"We're getting real close to where I say we call Palm Springs P.D. and cut them in on this, Sidney. We coulda *bought* it tonight, if that creature from the black lagoon turned on *us* instead a Billy Hightower."

"Another day, Otto. Let's see how it develops after *one* more day."

"One more day," Otto sighed. "Wonder if it's too late for room service. I think I got me a live one after all. Something in my stomach just did a two and a half forward somersault, with a full twist."

•

# LINES AND SHADOWS
•

*The true story only Wambaugh could tell. The media hailed them as heroes. Others denounced them as lawless renegades. They were the cops of the Border Crime Task Force, and it was their job to stop the ruthless bandits who robbed, raped and murdered the thousands of defenseless illegal aliens slipping into the U.S. Disguised as illegal aliens they walked into the violent shadows along the border and came close to the fragile line within themselves.*

Clouds like banks of foam blew in over the canyon mouth when Manny and Joe Castillo started in. Stunted trees with withered fingers pointed up and away from the canyon floor. Joe remembered the trees.

Manny and Joe walked about 250 yards along the creek bottom and soon they came to a curve in the creek where the trickle of polluted water snaked sideways and the brush grew thick. There seemed to be cloud shadow everywhere. Then from the twilight shadows a very ragged alien stepped from behind a hill of mesquite and stood silently staring at them. Then another man, this one twenty-three years old, the same as Joe Castillo, and wearing a creamy leather jacket, mocha slacks and boots. Joe admired the young man's clothes. There was never a pollo *or* bandit dressed like this. His left hand was down at his side. When he brought it up and extended it, they saw that his taste extended to firearms. He was holding a beautiful .45-caliber automatic pistol with silver grips. He was pointing it right at Manny Lopez's right eyebrow, which had leaped into a shocked and spiky interrogation point.

The two Barfers went instinctively to their haunches and tried to get into character, which wasn't easy. Joe Castillo customarily talked with his hands, long graceful fingers fluttering like bird wings. Ordinarily he was the world champ of body language. He hunched his shoulders, dipped his head, swayed his torso, squirmed his hips, always with the hands fluttering and gesturing. But not now. This was the first time in his young life that he had ever been face to face with a gun muzzle. Joe Castillo had turned to stone.

The gunman said, *"¡Migre!"* letting them know he was an immigration officer—from which country he didn't say.

The man held the gun in his left hand. He kept it just a few feet from the face of Manny Lopez. This was the third time a man representing himself to be a Mexican lawman had shoved a gun into the face of Manny Lopez. But this time Manny didn't pull a gun and badge and have a Mexican standoff. Not by a long shot. This time Manny had a very bad thought about himself slithering through his brain. The thought was this: You're gonna *die.*

The .45 was cocked. Then for some reason the dapper stranger moved the gun to his left and pointed it at the face of Joe Castillo, who squatted four or five inches to the right of his sergeant.

It was all happening so slowly that Manny Lopez couldn't believe it. It *is* like in the movies, he thought. Time *does* slow down. And then Manny stopped thinking that he was going to die and stopped thinking about time slowing down and stopped thinking about anything but the two-inch Smith & Wesson .38 in his shoulder holster.

While the .45 was aimed at the face of motionless Joe Castillo, who thought of inching his long fingers toward his own gun, Manny snatched the .38 from his holster and began jerking the trigger as it came up.

PLOOM PLOOM PLOOM PLOOM PLOOM! is the way it sounded in the ears of Joe Castillo. Then things speeded up for him as the dapper stranger began whirling, spinning, jerking. He was jerking back and forth like a wolf in a shotting gallery. Then Joe heard a BOP! as he saw the dressy dude going down.

The shot was from Joe's own gun and he found himself firing at the raggedy partner, who was flying across the creek bed, screaming his head off. Joe popped another cap and the raggedy partner went down.

The only transmission received by the frantic cover team of Dick Snider and Robbie Hurt was Joe Castillo yelling into the Handie-Talkie: "He's shot! We need cover!" which sent the Barfers running in all directions, mostly wrong.

•

# ECHOES IN THE DARKNESS

•

*Joseph Wambaugh's further probe of true crime took him out of L.A. and into a quiet Philadelphia suburb where a vicious murder would result in one of the most massive investigations in American history. Schoolteacher Susan Reinert was found wedged into her hatchback car in a hotel parking lot. Her two children had vanished. And the Main Line Murder Case burst upon the headlines—a Gothic tale of bright academics, upright citizens, and strange nocturnal habits.*

On Friday afternoon, Florence Reinert had the opportunity to speak on the phone with her former daughter-in-law. Susan intended to take the children with her to Allentown the next morning to a Parents Without Partners workshop.

That evening Michael was to play in a father-son softball game with the cub scouts. The game was being played at a church about half a mile from Susan Reinert's home. Ken Reinert arrived at the church with his second wife, Lynn, just before Susan showed up with Karen and Michael.

The game had lasted only a few innings when unexpected thunder sent everyone scurrying inside the church hall where the regular cub scout meeting was to be held.

Ken and Lynn sat in the back while the pack leaders tried to control thirty noisy kids. Suddenly Ken looked toward the doorway and his former wife was standing there. She was still dressed in a white knit blouse with multicolored stripes and blue jeans.

Ken was supposed to deliver Michael home after the game and wondered if something was wrong, but before he could ask she signaled to Michael, who ran to the back of the church hall, and they walked out together. Michael did not return.

Ken and his wife Lynn couldn't figure this one out, so they left for home at about 8:30 P.M. Fifteen minutes later there was a sudden cloudburst and Ken Reinert was standing on his front porch when the phone rang.

It was Michael. He told his father that he was sorry for leaving without an explanation, but that he had to get home to "scrub his floor" because they were going away.

His dad couldn't figure *that* one either because Michael had never scrubbed a floor in his life.

He said to his son, "Michael, where're you going?"

Michael called to his mother and said "Mom, Dad wants to know where I'm going."

And Ken heard his ex-wife say, "Well, why don't you tell him you're going bowling with PWP."

It was very strange these days for *all* the Reinerts. The children had become uncommunicative and evasive when it came to their mother's business.

That evening the president of the regional council of Parents Without Partners received a call from Susan Reinert, who said, "Something's come up. Something personal and I don't want to talk about it. Could you have someone cover for me at the Saturday workshop in Allentown?"

ECHOES IN THE DARKNESS copyright © 1987 by Joseph Wambaugh.

•

# THE BLOODING

•

*From the picture-perfect English village of Narborough comes a landmark in police detection: the first murders solved by genetic fingerprinting. A teenage girl is found murdered along a shady footpath and though a 150-person dragnet is launched, the case remains unsolved. Three years later the killer strikes again. To find the killer, it will take four years and the eventual "blooding" of more than four thousand men.*

By late December, after many members of the inquiry had voluntarily given up their Christmas holiday to work on old and new leads—and after the *Leicester Mercury* had printed a special four-page edition containing every salient fact and photo that might help the police, and shoved this edition into every letter box in the three villages—Supt. Tony Painter and all subordinates were required to suspend their disbelief. It was going to be assumed that genetic fingerprinting actually *worked*.

The ranking officers held a gaffers' meeting with DI's Derek Pearce and Mick Thomas. The subject was blood.

Chief Supt. David Baker said, "We're going to try something that's never been done."

Baker had sold his superiors on an idea—a campaign of voluntary blood testing for every male resident of the three villages. Anyone who'd been old enough to have murdered Lynda Mann in 1983, young enough to have produced the indications of a strong sperm count found in the Dawn Ashworth semen sample.

Both inspectors felt that Tony Painter was still convinced of the guilt of the kitchen porter. He'd wanted the Regional Crime Squad to do covert surveillance on the boy after his release from prison, but the police administration would not permit it. They knew that Tony Painter still kept the kitchen porter's file under lock and key.

But David Baker had apparently begun to believe in science. Alluding to the kitchen porter, he said that day, "He's either a co-conspirator or he's innocent."

By that they understood that Baker must have been pondering genetic fingerprinting. Regardless of what reservations any of them had over the guilt or innocence of the kitchen porter, or the efficacy of genetic fingerprinting, David Baker had decided that he was going to seek permission to do it, and he did.

In a compromise with his second-in-command, Baker never admitted publicly that the kitchen porter was probably innocent. His reasons for blood testing didn't mention Dawn's killer or killers. He kept it intentionally ambiguous so that everyone could save face.

He simply said to his two DI's, "Find the man who shed the semen."

The decision was made, and announced the day after New Year's 1987, that the murder inquiry was about to embark on a "revolutionary step" in the hunt for the killer of Lynda Mann and Dawn Ashworth. All unalibied male residents in the villages between the ages of seventeen and thirty-four years would be asked to submit blood and saliva samples voluntarily in order to "eliminate them" as suspects in the footpath murders.

The headline on the 2nd of January announced it:

### BLOOD TESTS FOR 2,000 IN KILLER HUNT

As several members of the inquiry later said, "We *had* to have blind faith in genetic fingerprinting."

•

# THE GOLDEN ORANGE

•

*When forty-year-old cop Winnie Farlowe lost his shield, he lost the only protection he ever had. Ever since, he's been fighting a bad back, fighting the bottle, fighting his conscience. But now he's in for a special fight. Never before has he come up against anyone like Tess Binder.*

When he got back to the bar there was a blonde in an ivory cardigan, with navy blue and ivory striped pants, sitting at the bar. She wore a funnel-necked navy pullover under the cardigan. The outfit looked nautical without any of those corny little anchors.

Winnie tried not to stare. Her boyfriend or husband might be in the john. Another wealthy couple out slumming, probably. He figured her for the old yacht club. She was understated but elegant. Tiny ear studs and a platinum wristwatch with art deco dials, that was it for the jewelry. Not even rings on those long elegant pampered fingers. Not even a wedding ring!

Her hair was the color of melted butterscotch, streaked with golden highlights, like that last halo before sunrise he used to see from his boat when he was bobbing level with the horizon. Before that bitch Tammy took his sloop and anything else with a salvage value.

The woman at the bar might as well have been wrapped in razor wire, she looked so unapproachable. She smoked, and sipped something that looked like an Americano. Winnie knew he was blasted when he heard himself blurt, "I used to have a twenty-nine-foot sloop. She was Danish, a double-ender with a canoe stern. A production boat, glass, but sweet. Once I was sitting in her cockpit at four A.M. And . . . Do you know that water boils at a higher temperature at sea level?"

"I only know about water in my kitchen," she said. Smiling!

"Anyways," Winnie continued, "the coffee was very hot. The sun was beyond the curvature of the earth but getting ready to rise at the stern. I put my coffee down and watched this light start at the horizon. And there were these cottonball clouds so heavy you couldn't star-sight. The clouds were so full in that breeze, well, the fan spread and it sorta backlit the clouds. That's the way it *must* a looked before there were continents. The light, it was something like . . . it was like the color a your hair. Well, I jist wanted to tell you that."

She was *really* smiling now, not the way a beautiful woman usually smiles at a drunk. He had lots of experience in such matters. She smiled like she meant it, with those wide vermilion lips of hers!

"My name's Tess Binder." She held out her hand and he took it. She was strong.

"I'm a sucker for women that shake hands like a man," he said.

She chuckled. Like wind chimes! She said, "That's flattering. I guess."

# FUGITIVE NIGHTS

*Returning to glamorous Palm Springs and its ever-mysterious desert, Joseph Wambaugh delivers a suspense novel rich in astonishing plot twists, fascinating characters, and the penetrating humor for which he is famous. Lynn Cutter—cynical, hard-drinking, and soon-to-be ex-cop—is enlisted by already ex-cop and struggling private detective Breda Burrows to help on a strange case that requires the one thing Lynn has that Breda doesn't . . . and leads them both down a shadowy path and on a collision course with another cop and his obsession with a dangerous fugitive. . . .*

"So Mrs. Devon," the P.I. said, thinking there'd been enough small talk. "How can I help you?"

"It's about my husband, Clive," Rhonda Devon said. "I'd like you to follow him."

That was a bad start. The P.I. never had any luck with people named Clive or Graham or Montgomery, and once had served at Hollywood detectives under a captain named Clive, hating his guts.

"I think he's preparing to have a child. And I can't understand why. I don't care if he has one mistress or ten! But he's having a *child*. I have to understand that."

"How old is your husband?"

"Sixty-three."

"And how old're you, if I may ask?"

"Forty-four. I've never had any children, and as of last December I won't be having any. I went through the change rather early just like my mother and both my sisters. Clive's obviously planning to have a child by a surrogate! Perhaps he's planning to leave me!"

"Do you care?"

"Yes, very much."

"Did he make you sign a prenuptial?"

"No."

"Then you stand to inherit when he dies?"

"Oh, yes. We've been married for thirteen years. He can't legally leave all his money to a new wife and child."

"Well, did you ever want children?"

"No, nor did he. Neither of us had happy childhoods so we thought we'd keep our neuroses to ourselves and not pass them on."

"Mrs. Devon, why don't you just *ask* him why he made this little deposit that's driving you nutty?"

"Oh, I'd never pry. Nor would he if the roles were reversed. We're each very independent. We live apart a good deal of the time. I prefer our main house in Beverly Hills and only come here two weekends a month. He stays here all the time, even in the summer. I seldom can get him to spend forty-eight hours at our other home."

"Do you and your husband still . . ."

"He had a cardiac bypass. Arterial insufficiency allows him to ejaculate, but he can't get an erection. We haven't had sex for about five years." Then she added, "At least together."

FUGITIVE NIGHTS copyright © 1992 by Joseph Wambaugh.

Here is a look at Joseph Wambaugh's
novel, *Floaters*, available in
paperback from Bantam Books

# FLOATERS

# by
# Joseph Wambaugh

"Can you imagine how my life changed?" Ambrose Lutterworth asked Blaze for the third time, if she was counting.

"I can only guess," she said, deciding not to accept any more wine. She was getting shit-faced.

"I've traveled the world, not as a tourist but as the Keeper of the America's Cup. I've met *kings*. Sometime I'll tell you about Princess Anne. She was the loveliest person. Not regal, a real person. I found Prince Rainier to be regal, though."

She had just enough alcohol boiling in her belly that she was getting irritable, something she tried to avoid with clients. "Let's talk *business*, Ambrose. How about it?"

"I want to *remain* Keeper of the Cup," he said. "I don't want it to end. Not yet."

"That's talking business?"

She plumped up a throw pillow behind her back. A sofa spring was on the verge of breaking through the fabric. She wanted to go home.

"Pour yourself another glass," he said. "I'll be right back."

Against her better judgment she poured half a glass, emptying the second bottle. Blaze figured he'd gone to the can. Guys his age, they were *always* running to the can. Prostrate problems, they said, as if she didn't know. She'd massaged a *lot* of prostates in her time. Blaze Duvall figured she could be a pretty fair urologist if handling prostates had anything to do with it. Most of her clients expressed admiration for her long, graceful fingers. Of course she kept her nails clipped short.

When Ambrose returned, he had a folder full of papers, photos, and clippings. He opened it on the coffee table.

"See this," he said, pointing to a newspaper photo of a sailboat crunched on the ground.

"Yeah?"

"That's an America's Cup boat. Belonged to the French, who also had problems with their backup boat. That other one lost a keel and rolled over like a harpooned whale."

"So?" Blaze looked at the photo, then back at Ambrose, who at last had loosened his tie.

"I don't know for sure who's going to be the defender, but I know for sure who's going to be the challenger: New Zealand. The Kiwis. And they're the opposite of the French syndicate. All business. Ruthlessly efficient and professional. They've got two fast boats. And no American defender is going to beat *one* of those boats."

"You don't say."

"The Kiwis have NZL thirty-two and NZL thirty-eight. In nineteen eighty-seven they won thirty-eight victories to only one defeat through the trials, yet they ended up losing four races to one to Dennis Conner in *Stars and Stripes.* In ninety-two the Kiwis were one win away from the challenger trophy, yet they lost four straight to the Italians. This time they're hungry and they vow it won't happen."

"Okay, Ambrose," Blaze said, her patience gone. "Our business deal has something to do with the America's Cup. What the hell is it?"

"It's this. The Kiwis' thirty-two boat is better, *much* better than their thirty-eight boat. The defender will have no chance against the thirty-two boat. But we'd have a chance, a good chance in my opinion, against the thirty-eight boat. I've done my homework. I'm well-enough connected to have gathered good intelligence. I feel in my gut that the thirty-eight boat can be beaten."

"And what do you expect me to do? Exactly what?"

"I want you to help me. It's not personally risky, mind you, but I want nothing less than the destruction of the thirty-two boat. They'll have to race the thirty-eight in the finals. I think our defender can beat the thirty-eight."

"And how would I be able to help you wreck a boat?" Blaze asked. The guy was loony! A loony old geek whose life revolved around a dumb trophy.

"The Kiwis have seven people in their syndicate who they call designers," Ambrose continued. "Sail and hull designers, appendage designers who crafted their keel, and analytical designers. They have a meteorologist. They stop at nothing to ensure that all the people in their syndicate are loyal, dependable, dedicated. They even brought their own crane operator with them."

Despite her cynicism, Blaze was getting slightly interested.

He looked so serious, and he was cold sober, unlike her. "They must have security people guarding those boats," she said.

"The Italians had *fifteen* last time. And a dozen TV monitors. Even dogs. The Kiwis have only two men, but they're police officers. Real police officers. Brought them all the way from Auckland on leave from the New Zealand Police. They're well protected in their compound."

"I hope you're not going to say you think I can get to one of *them*?"

"Impossible," Ambrose said. "Those people have national pride in winning the Cup that Americans can't even imagine. Auckland's called the City of Sails because they have more sailboats than cars. There's half an hour of live coverage on their major television channel every night during the challenger trials alone. But there's a weak spot in their program. In every program. A boat can simply be dropped when it's being lifted in or out of the water, and the lifting happens almost every day. Their boat can be dropped just like the French boat was dropped. They're loaded into the water in basically the same fashion, either by crane or by travel-lift. A crane operator can make a mistake. It happened to the French, it can happen to the Kiwis."

"I'm not much at operating cranes," Blaze said. She felt like saying the only machines she could work were electric: a toothbrush and a dildo, which she used on her clients, not on herself. Instead she added, "You want to bribe the guy that does the lifting, is that it?"

Ambrose smiled. "You truly are a bright young woman, Blaze. You're on the right track."

"What? *Tell* me, Ambrose!"

"I want to . . . *incapacitate* the New Zealand crane operator who runs the travel-lift. I want it to happen on the last day of the challenger trials when they're racing the Aussies. When they're on the verge of finishing off the competition. They'll have to replace their man without notice. They'll be forced to turn to the boatyard they rent their space from."

Blaze tried to keep her mouth shut. This guy was so anal, he had to get around to everything in his own time, but she had to ask. "Do you *know* the boatyard guy?"

Ambrose nodded. "There're three crane operators working there, but one of them is the brother of an American woman who's married to a Kiwi sailor. He'll be the one they'll go to on such short notice because his brother-in-law's a New Zealander and because he's very experienced and worked for racing syndicates in the last America's Cup regatta. I used to be a client of that boatyard. He's hauled out my sailboat many times. I *know* that man will be the one who gets the job."

"You're saying that something's gonna happen to the New Zealand crane operator."

"Yes."

"Like what?"

"I'd like you to meet him. I know where he and all the Kiwis will be this Thursday evening. Where they are every Thursday evening: at the AC/DC party."

"What's that?"

"The America's Cup Drinking Club. A different bar in town hosts a party once a week. Nobody knows where it'll be until the morning of the party, when the organizer sends a fax to each syndicate. The crane operator will be there, and if you accept my proposition you'll be there, too. He'll leap at the chance to have a drink with a girl like you. Who wouldn't?"

"And then?"

"Nothing *yet.* You have drinks. You get acquainted. You become friends. The important thing is, you'll also be wherever he is the night before they're to clinch the challenger series."

"What would I do to . . . *incapacitate* him?"

"You'll put some medication in his Steinlager."

"In his what?"

"It's the New Zealand beer that sponsors them. Their holy water. They all drink it. The drug is something I've kept since my mother's last days. It won't do him any real harm, but he won't be in shape to go up on a travel-lift the next morning. The Kiwis will be panicked. They'll have to call for help."

"You plan to bribe the substitute crane guy, is that it?"

"I'm hoping *you'll* take care of that. That's what the business proposition is all about. Making a deal with Simon Cooke, the crane operator."

"Why me?"

"I know Simon Cooke. Loves women, loves to drink, loves to go to Tijuana and gamble on the jai alai. Loves to talk. He's a perfect candidate to make a deal with a beautiful girl. After he gets to trust you."

"Wait a minute!" Blaze said, more soberly. "You want me to get next to this guy Simon? And get to know him? I think I know what *that* means. And then ask if he'll drop the New Zealand boat? Drop it on the ground?" She sat up, staring at the picture of the French sloop with its keel poking through the hull.

"Yes," Ambrose said. "For ten thousand dollars. That's a lot of tax-free cash for a guy who makes fifteen dollars an hour. I know he'll do it."

"How many jobs can he get *after* he drops a boat?"

"He'll think of something to blame it on. An excuse as to why it wasn't his fault. Nobody can ever prove anything when things like that happen."

"Why don't *you* make the guy the offer?"

"I don't dare get anywhere near this. Do you know what would happen to me if I got connected to a plot to sabotage a challenger's boat?"

"Yeah," Blaze said. "Same thing that'd happen to me. You'd go to jail."

"That's the least of it," Ambrose said. "My reputation—my life—would be . . . *gone*. I don't like to think about it. No, I can't be directly linked to Simon Cooke. Nobody must ever know about me."

"And what do I get outta this . . . business proposition?"

"Just about everything I have in the world," Ambrose Lutterworth said. "Fifteen thousand dollars. My life savings. My annuity, you might say. You get it all, if you persuade Simon Cooke to do it. And *if* he does it."

"And you get . . ."

"The Cup. I get to be Keeper of the Cup for another four years at least. Who knows? Maybe for a lot longer."

"This is pretty nutty," Blaze said. "I gotta think about this."

"There isn't much time," Ambrose said. "The last race between the Kiwis and the Aussies is only three weeks away. There's a lot to do before then."

Blaze said, "Let's say I could give the New Zealand guy his sleeping pill. How do you know for sure they'd call Simon Cooke instead of somebody else? And what if he weasels out? What do I get for trying?"

"You'll get five thousand, whether or not Simon bites. Whether or not he does the job. You know all about me now. You can trust me just as I'll have to trust you. I know Simon won't turn you down. I've done my homework, Blaze. This will *work*!"

"Why'd you pick me, Ambrose?"

"I've been waiting," he said, "for misfortune to strike the Kiwis like it's struck everyone else. A boat has been dropped.

Another sunk. A keel fell off after being hit by a rogue wave. A mini-tornado even struck one of the compounds. An aircraft carrier almost cruised into the racecourse one foggy day. But nothing happens to the goddamn New Zealand boats! I can't afford to wait any longer. Something has to be done. The idea came to me a few days ago."

"Why *me*!"

"Because," he said, "you're smart and beautiful and discreet. And you're the only person I know—the only person I've ever known in my entire life—who works outside the law."

"What I do is a misdemeanor if I'm caught," she said. "What you're suggesting is a heavy-duty felony."

"It's only a matter of degree," he said. "There's nobody else in my life who can do it."

"I'm going to sleep on it," she said. "And I get five hundred for tonight. Right?"

"Of course. But I was wondering."

"Wondering what?"

"If you could give me a quick . . . massage?"

"Okay," Blaze said. "In the bedroom?"

"Did you bring the warming cream?" Ambrose wanted to know.

Ten minutes later Ambrose Lutterworth was lying naked on the two large beach towels that Blaze Duvall had spread on his queen-size bed. She was standing beside the bed, squeezing some Icy Hot on her palms. She was naked except for black bikini panties. Blaze smiled professionally when she spread the cream over his buttocks, kneading the muscles gently.

"That's wonderful, Blaze!" he said. "Just wonderful! You have splendid hands!"

Blaze glanced into her bag, fearing she'd forgotten the condoms, but no, a package was lying there, along with the toys that clients requested: a feather for tickling their balls, a vibrating dildo for rectal stimulation. Toys.

"Turn over, darling," she said, trying to speed things up so she could go home and think.

"No, I don't need it this time," Ambrose said. "Just rub on some more cream, please."

So at least she wouldn't have to blow the crazy old bastard.

While she was rubbing in the Icy Hot, careful to avoid tender tissue, he said, "Blaze, move the lamp a bit to the right, please."

She did it and saw that he wanted light shining on a framed

photo on his dresser. In the photo Ambrose was standing by a sunny foreign harbor with a young woman in a white dress.

"Cap d'Antibes," he explained. "She was just a girl I saw by the waterfront and I asked if she'd pose for a picture with me. Are you at all familiar with the South of France?"

"No," Blaze said, working his right buttock so strenuously that he grunted in delight, finding her as sultry as a cheetah—rubbing, purring, blowing her warm breath on him.

Then he said, "It's between Nice and Cannes. After I got the picture taken I went up to my hotel room and sunbathed nude on the balcony. No problem is the people across the courtyard could see me. In the South of France nobody worries about such things." Then he said, "Blaze, I'd like to turn on my side, but please don't stop."

She was working up a sweat from the wine. Beads of heat lay on her upper lip, her mouth brooding and sensual.

She paused to let him turn on his side and saw his watery blue eyes gaze up at the photo, his brows silver-flecked in the lamplight. When she began massaging again, he said, "The Cup was on the balcony with me as I sunbathed that day. I put it on a chair and watched it. The sun glinted off the silver and the sun's rays were hot, *very* hot, reflected onto my bare bottom. I didn't care if I got burned. I didn't care about anything. I don't think I've ever felt so at peace with myself. So contented with my life. So . . . blissfully happy."

Then Ambrose Lutterworth surprised Blaze Duvall by reaching down and slowly stroking his penis.

Blaze smiled encouragingly, but he never looked at her. Never stopped gazing at the picture. In just a moment he was erect, and he didn't take his eyes from the photo until he was through.

This takes the cake, Blaze thought, watching Ambrose Lutterworth reliving an extraordinary moment in his life: when he'd sunbathed on a hotel balcony in Cap d'Antibes, literally basking in the reflected glow of the oldest sporting trophy on earth: The America's Cup.

Or, as Blaze later explained it to Dawn Coyote, "I got five hundred scoots to watch this geek skipping down memory lane and slapping old Porky. While I set his ass on fire."

CPSIA information can be obtained at www.ICGtesting.com
Printed in the USA
LVOW060709180812

294834LV00001B/34/A